D0295893
Llyfrgelloedd Caerdydd
www.caerdydd.gov.uk/llyfrgelloedd
Cardiff Libraries
www.cardiff.gov.uk/libraries
ACC. No: 05150137

The Hard Way

The Hard Way

ADAPT, SURVIVE AND WIN

Mark 'Billy' Billingham

with Conor Woodman

SIMON &
SCHUSTER

London · New York · Sydney · Toronto · New Delhi

A CBS COMPANY

First published in Great Britain by Simon & Schuster UK Ltd, 2019
A CBS COMPANY

1 3 5 7 9 10 8 6 4 2

Simon & Schuster UK Ltd
1st Floor
222 Gray's Inn Road
London WC1X 8HB

www.simonandschuster.co.uk
www.simonandschuster.com.au
www.simonandschuster.co.in

Simon & Schuster Australia, Sydney
Simon & Schuster India, New Delhi

A CIP catalogue record for this book
is available from the British Library

Hardback ISBN: 978-1-4711-8673-8
Trade Paperback ISBN: 978-1-4711-8674-5
eBook ISBN: 978-1-4711-8675-2

Typeset in Minion by M Rules

Printed and bound by CPI Group (UK) Ltd, Croydon, CR0 4YY

CONTENTS

The Hard Way

PROLOGUE

I had now been in 3 PARA for a few years, so I felt like I could handle pretty much any test thrown at me. Having gone up for a promotion to be part of a recce troop within the battalion, I was warned this phase of training in the jungle could be the toughest test I'd ever faced. Here I was, at 0500 hours, ready to ship out via helicopter, in a scene from a Vietnam war movie. Six big UH-1s were lined up, blades spinning, the sound deafening us, as my unit ran over the ground and loaded into them one by one. When we landed, we would be split into our assigned patrols of eight plus a member of the DS (Directing Staff) – our boss for the exercise. The helicopter took off and, suddenly, there I was, flying over the jungle canopy, like a field of steaming broccoli below. I thought to myself how fucking cool it was. It really was something else, a whole different league to anything that I'd experienced before. I was flying in a helicopter over an Asian jungle. This was everything I'd dreamed of, and now here it was, coming true.

The helicopter set us down in a small clearing surrounded by hundreds of miles of forest on all sides. We got off and began unloading the kit. As the helicopter lifted off again, the noise of the blades faded gradually, only to be replaced by the sound of the DS. He stood in front of us, a big old straggly beard on him, looking rough as fuck, dirty face, clothes already soaking wet with sweat and carrying his weapon. He looked exactly like the real deal, like he'd just come straight off a tour of duty.

He told us to follow, and we carried the kit to the area that would be our camp. That first day we spent building, slinging our hammocks, learning the layout. We cut a path down to a wash point on the river, where we would wash from now on, and a water point, where we would only fill water upstream of the wash point, of course, so you don't drink people's piss. There was one toilet area, and the DS explained all the rules and regulations of that, too – don't throw any rubbish in the fucking toilet, which people have done, trying to get rid of stuff they didn't want to carry. He explained that, while we were in the jungle, we would leave nothing behind, just as we would on operations – nothing that the enemy could use to prove we had ever been there. It was the first sign that this was going to be real; everything we did, we did as though we were in action.

'So, any of you drop anything in there and you'll fucking get in to get it out again,' he warned.

For the next days, we were constantly on patrol. Each morning, the DS would give us co-ordinates for where we were and where we were expected to go. We moved as though we were on patrol, carrying full kit, moving silently, weapons ready, navigating while looking for the enemy. Moving through the jungle is hard. The terrain and the trees meant that a kilometre of jungle could take a few hours to navigate through. We'd break the ground in front of us into 10-metre increments, each 10 metres equivalent to seven lots of two strides. All the while we patrolled, counting, scanning the trees, maintaining a bearing, keeping silent, knowing that, somewhere hidden in the tree line, the DS was watching, evaluating our every move. It was all slow and methodical and time-consuming; and, because of the slowness, the measured pace, the type of terrain, the equipment we were carrying, the pressure and the fact that we were sweating the whole time, it was mentally exhausting.

We were limited to what we could carry, and everything felt heavy in the draining humidity. I had one set of dry kit and one set of wet kit, and it was vital to keep it that way. At the end of each day, as the darkness set in and everything settled down, I changed out of my wet kit and put on my full set of dry kit to sleep. My wet kit went into a plastic bag, ready to be worn again the following morning. I slept next to my rifle, and my boots sat upturned on sticks to stop nasty creatures climbing into them or the rain getting them wet in the night. Every moment we were primed, always ready to fight or run. We never slept without footwear of some sort. Every morning began forty minutes before it got light. In the pitch dark, I took out that wet kit, still soaking, sodden with yesterday's sweat and stinking like death, and put it on again. My dry kit went back into the bottom of my Bergen, safe for the next night's sleep, and the day's patrol began just like the day before. Relentless, repetitive, exhausting.

The patrols were soul-destroying. All of us felt uncomfortable, carrying so much weight, constantly tired and hungry. All the skill sets we'd been taught were being tested over and over again as we'd stop and get down into a fire position every 20 or 30 metres. In that environment, it's vital that the Bergen is taken off slowly and put down gently, so as not to make any noise, but doing that fifty times a day is a real test of character. The DS were watching, looking for signs of tiredness and weakness, looking for the person who threw their Bergen down or even dropped it. They were looking at how we held ourselves together. How we were working as individuals, but also how we were working as a team. So, if one person was taking off the Bergen and slowly putting it down, the other was still, watching, covering. Then, once the first Bergen was down, it became your turn to cover for me, and I took my Bergen off. It was those small things that the DS looked for to understand the characteristics of the team and of the person.

We got no feedback. The DS just silently took notes, evaluating us but never sharing their thoughts. That played havoc with my mind every night in my hammock, as I went over the events of the day in minute detail. Did I mess that up? Did I make that mistake? Did anyone notice me do that wrong? Every night was a constant exercise in beating myself up and playing mind games in my own head as I imagined getting to the end of this hell only to be told that I'd failed for dropping my Bergen on day three or missing a checkpoint on day five. I realised what an idiot I'd been to ever think this was going to be easy. I'd massively underestimated the scale of the challenge. This was not what I was expecting and I'd actually made it harder still by coming in with completely the wrong attitude. I knew I had to work twice as hard to get my head straight and catch up.

Every morning began at a small hut on the edge of camp, which we called the schoolhouse. Before the DS briefed us on the day's patrol, he asked calmly and without judgement if anyone wanted to leave. I knew I was thinking the same as everyone else, that I'd love to put my hand up and just go home, but nobody wanted to be the first to do it. It was just like P Company, only in a hot climate. Then it happened. By day four, we all looked like we'd already been out there for a month. I glanced around the schoolhouse at all the beards and straggly hair and almost chuckled to myself at how rough we were looking. The men were filthy, gaunt and drawn, and we were staring at each other in a state of shock, searching for some sort of recognition from each other of how hard it was. The DS came in as usual.

'Morning, fellas,' he said. I liked how he always called us that, like we were his mates. 'Anybody want to get on the freedom bird to civilisation?'

That was the carrot they dangled in front of us. A short, two-hour flight away were white-sand beaches, cold beer and hot food. If you withdrew that morning, you could be on the beach by the afternoon. But I wasn't going for it. Sitting on my arse with a cool beer wasn't the sort of 'going a little further' that I had in mind. Just out of the corner of my eye, I saw a hand go up. I was stunned. I couldn't believe it and, at the same time, it lifted a weight from my shoulders. It was as though it was now a hundred times easier for me to quit. I wouldn't be the first. If he was quitting, then there was no shame in me quitting, too. But something inside me just wouldn't let me raise my hand. Even when another and then another hand shot up. Even when five men sat around me, hands raised, ready to get out of this torture and do something else with their lives, I couldn't do it.

'Okay, fellas. Just pick up your stuff. If you want to dress off to the back of the schoolhouse and go up onto the landing zone, somebody'll meet you up there.'

Already, the chief instructor was waiting by the helicopter, seemingly knowing that today was the day somebody was going to quit. I watched as five strong men left quietly, feeling a boost, somehow strengthened by their departure, because, if they couldn't hack it and I could, then maybe I was stronger than I'd previously thought. I sat still and waited for the DS to return his attention to those who remained and give us the day's task.

As the numbers thinned, the patrols got harder because there were fewer people to carry the kit. The radios and batteries, the heavy gun and all the extra ammo now needed to be shared between three or four rather than eight. Added to that, the people who left were missed. By now, some of those guys had begun to feel like friends, people who lifted the morale in the group or maybe made me laugh. As the weeks passed, I felt their absence, but I knew I had to be selfish. I had to focus on being the one to lift morale in their place. I had to rely on myself to maintain my own spirits. There were nights I went to my hammock with a genuine sense of pride. I could feel myself improving, getting stronger, sharper, and I was even a bit impressed with myself. It was a weird feeling, but also a nice end to each day to pat myself on the back and give myself a 'well done'. It helped me to keep going and remember that another day done meant another day I was still there.

Time passes slowly in the jungle. At first, I found myself checking off the days, but soon I was ticking off the hours. I was starting to wonder if I'd just pass out and fall over, because I constantly felt tired and dizzy, like I couldn't take one more step. The food was all dehydrated rations and we ate it cold most of the time, because fires can give away your location. On the rare occasions they let us eat a proper meal, I couldn't finish it anyway, because my stomach had shrunk so much that I felt full after a couple of bites and then spent the next day shitting through the eye of a needle. I became good at setting up my hammock so that it lined up with a window in the canopy, and I spent my time before I fell asleep looking at the stars, wondering what the kids might be up to back home. But thinking of home too much was also dangerous. The mental test was hard enough without adding an extra emotional layer to it.

We did get letters, even though we didn't really want them. There was nothing we could do if they contained bad news and none of us wanted to be dragging that around with us. There were stories of people who had received letters and immediately dropped out the same day. I had two letters in total. One was from Julie and the kids saying how much they missed me and describing all the stuff they'd been doing since I'd been gone, which was really nice. It made me sad for a moment, but then I thought about how I was going to make them proud and that actually picked me up a bit. In the final phase of the jungle exercise, the DS explained to our patrol that enemy patrols had been identified and were active in the area, harassing and assaulting local farmers and other civilians. Our mission was to, firstly, identify exactly where they were and who they were, with a view to either arresting or taking them on in a fire fight. We set off, around 15–20km away from our target, which guaranteed that we had to scale the highest feature, a 1,200ft hill, deep inside the jungle, to reach the other side. After three days and nights of slow, methodical progress, we found the area where we would begin searching for our targets. All the tactics of how to locate and track people were being put to the test and, again, silently, invisibly, we were being evaluated by the DS, watching from behind the tree line. The enemy patrols were being role-played by staff from the camp and, once we'd identified them and signaled back to headquarters, we were told to place them under close observation, do a close-target recce and see whether we could take them all out.

That night, I lay watching the enemy camp, waiting for dawn to come, when the attack would happen. It was howling down with rain all night, so the jungle floor was covered in leeches, there was shit everywhere, but I just didn't give a fuck. I was watching the little campfires, listening to the people's voices, and, suddenly, I felt something I'd never allowed myself to feel before: *This could be real.* Finally, we were ready to make our attack and, with the enemy now replaced with metal targets by the DS in the night, we mounted a fully armed assault on their camp, using live ammo. We captured the enemy targets and received the order to withdraw. Our patrol, plus all the other patrols that had made it to this point, began the march back, 3km out of the jungle, which we had to navigate, with the prisoners and all the remaining kit, to a helicopter landing site, for which we had only co-ordinates.

That last march was so arduous that it forced people to voluntarily withdraw, falling at the final hurdle. Fortunately, nobody from our group quit and, as we centred on the landing zone, the helicopters came in to pick us up and take us back to the base we'd left four weeks before. That was it, the end of the exercise, the end of jungle, the end of the hardest month of my life (until I requested selection). As we loaded up onto the helicopter, my DS, a little Scottish guy called John, looked at me.

'Billingham?' he said. I could barely lift my head to look back at him as he pointed at my face. 'There's no way on God's earth that you are getting into my army . . .'

I was absolutely devastated. I looked back at him, dazed, thinking, *Fuck. I've failed. After all that.* But then I realised he hadn't finished what he was saying.

'... with a beard like that.'

He roared with laughter and kicked my arse as I fell into the helicopter, and he jumped in next to us as we lifted up over the trees. I slumped in my seat, looking out over the jungle again, feeling so different now to how I had on the way in, wondering how it was that this had been my home for the past few weeks. An amazing feeling of relief rushed over me because, finally, it was done. The wind began to dry my uniform, and the fabric disintegrated as though it had been burnt by fire; dry, crumpling like paper on my skin. I looked across to one of the lads sitting opposite me. We both managed a half-smile at each other and, once again, I felt that rush of pride. I was absolutely hanging out of my arse, exhausted out of my mind, but I could find enough strength to muster the thought that this was it, this was everything to me. This feeling of accomplishment was all I ever wanted, why I joined the army, why I wanted to be a soldier – and it felt pretty fucking great.

INTRODUCTION

I should have died several times in action during my military career, I make no bones about that. It could have happened on the frontline, in a theatre of war; or going house to house fighting terrorists and insurgents, one of them getting the drop on me; or being captured by the enemy and interrogated, not knowing how it was going to end. There have been countless times when I've nearly lost my life. I have even been shot at, from point-blank range, and yet the bullets deflected off my equipment and I survived. There isn't a day that goes by when I don't think to myself, *I am lucky to be here.* A lot of my friends weren't so fortunate. The zones of war and conflict I have operated in for the SAS are still classified for obvious reasons connected to where I was, who I was with, and whom I fought against, but, for seventeen years of my life, the British Army would pay me to travel to the most dangerous places on the planet. They put me and others like me – my brothers in arms, if you will – in a succession of stressful and demanding situations. The army sends in the SAS because we are discreet. That discretion continues, and I intend to respect that. I am humbled and extremely proud of my military career and the fact that I rose to the rank of sergeant major (a WO1, to be exact) in the Special Air Service. I would ultimately be honoured by Her Majesty the Queen with an MBE for services to my country in the line of duty, and nothing, other than the love of my family, beats that.

I am also fortunate that, since I left the army, opportunities have opened up in my life, allowing me to pass on just a small portion of the philosophy that moulds what a special forces solider needs to have in his psychological armoury in order to be the best he possibly can, as well as survive. That brings a certain level of public recognition, for which I am grateful, but I still like to think I promote the best ethics of the Regiment. One of the reasons the SAS is such a world-renowned organisation is that, outside of their base in Hereford, they are still an enigma. People might think they know what we're about, but, unless you have served, you don't have a fucking clue.

I want *The Hard Way* to offer the reader a chance to look at my life and see how I overcame hurdles, both physical and mental, in order to achieve my goals. It hasn't been easy at all – at times, fucking desperate – but this is the life I have chosen, and I am proud of what I have achieved.

But the story isn't finished.

'Always a little further.'

Life Lessons I Adhere To

For a kid growing up in the West Midlands, being streetwise came naturally to me. From a young age, I immediately gravitated towards the older generations and I have had the benefit of being influenced by many of these people. They helped me to form my mantra: 'Always a little further.' This has been so typical of my life, through all my hardships, hence the reason this book, *The Hard Way*, was written. Even though the literary world is alien to me, and to be commissioned to write my memoir was a massive opportunity, I embraced it in order to push myself. You, too, should always want to try to go that little bit further.

I am under no illusions. I know I have been privileged to have had an incredible support system, and I have met people who have invested in me throughout my life. I know many of you out there are not so privileged. And, for that reason, I want to share the lessons learned along my journey, in the hope that I can help you all. In no way do I feel I am some sort of 'life expert', and I don't portray myself as that when I give talks around the country; however, my journey through life has proven that, against all odds, amazing results and achievements can be reached.

1. One of the biggest questions that you may have on your mind if you're ambitious is: how did I get to where I got to and how could you do the same sort of thing in whichever field you want to succeed in? If you don't reach your goal, you will still be in a better place than when you started, as long as you try. This I always say because, when I first joined the Parachute Regiment as a young 17-year-old boy, I was not only the youngest, but also one of the skinniest of all the other recruits. Even though physically I was the most unlikely to succeed, I continued to remind myself that, if I just tried to get a little further ahead each day, I would be in a better place and might even eventually reach my ultimate goal. Which I did, little by little . . . and so can you.

2. When I was serving in elite units, I'd receive regular reports on my performance, like everyone else. They were generally a mixed bunch – comments like, 'Billingham always sails too close to the wind,' or, 'If Billingham spent as much time

on the battlefield as he did in the bar, he'd be the best soldier in the army.' It seemed unlikely for a while that I'd ever progress, and it would have been easy to believe those reports and settle, but I didn't. Instead, I sorted my shit out and won the respect that earned me promotion and senior roles. It was an important lesson for me to learn: it doesn't matter what other people say about you, because, at the end of the day, the only thing that matters is what you think about yourself. Other people will try to hold you back, but, if a working-class lad who came from nothing can realise his dream to lead elite soldiers into battle, anything is possible.

3. As an SAS solider, I never knew if I would walk away in one piece from dangerous situations. However, I found that belief, courage and conviction can help you navigate the dangerous situations best. It is your job to come up with new options, new ideas, and, against all odds, new solutions to each problem that creeps up in your daily life. I have been tasked with roles where I was making crisis-management decisions, often in life-or-death scenarios. My primary task was to keep the people trusted to my care alive. This

required operating at a higher level than I was used to, but it helped me learn what my own physical and mental limitations were. Challenging yourself to take on greater tasks can help you overcome hurdles that you have created in your mind – whether through lack of confidence or because you failed the first time. Always keep trying.

4. When everyone is bigger and stronger than you and looks the part, have the self-assurance to know that your greatest weapon is your experience. I remember a situation with Clint Eastwood when all the bodyguards on a movie set were twice my size, more intimidating and better typifying the image of what a bodyguard should look like. As the anomaly, I was surprised to be appraised by Clint himself, who sidled up to me on his film set, stating that he was intrigued to know what my speciality was. 'Do you do karate? Or taekwondo?' he asked. To which I replied, 'Clint, my real weapon is my mind and my experience. I think. I plan. I execute. If I am rolling on the floor with an assailant, I haven't done my job properly as I should have neutralised the threat before it came to the point of that situation

arising.' He raised his eyebrows, smiled and walked away. But I was only telling him the truth. It is these skills that have helped me advance in my life and my career.

5. There's a saying in the SAS: 'You're only as good as your last job.' There was never any time for self-gratification. Time was spent wisely on lessons learned. What could we have done better? What could have gone wrong? We analyse our mistakes and aim never to repeat them, only to be better. Every day is a school day. Listen, learn and pass on your knowledge and experience, because those you invest in are the future. That is the SAS way.

6. I witnessed some of the worst conflicts and natural disasters known to man or woman. You will read about some of them in the pages of this book. Those experiences gave me the platform I stand on today, which is to always go a little further in everything I believe in. I know we all have the power to make the world a better place. I think it is our duty to help more, give more, share more. During my darkest days, it's humbling to remember that there are others in

worse situations, and usually it's through no fault of their own. Bad governments, bad people, bad decisions, bad luck. It reminds me to get out of my own head and do the right thing for the greater good.

7. When I'm appearing on the Channel 4 series *SAS: Who Dares Wins*, I am a ruthless bastard. I'm not there to please anybody; I'm not there to make friends; I'm there to give them a brief taste of what SAS selection would be like. There will be times in your life when you may be criticised; you may be unjustly accused of something; you may piss a lot of people off because of your decisions. Unfortunately, that's the way life goes, and, if you want to please everybody, you're on a certain road to failure. This is especially true when it comes to being successful. As soon as you put your head above the crowd and say, 'I'm going to do this,' or, 'I'm going to go for that,' then you're going to have certain people criticise you and doubt you. Some people, like your family, may do it with your best interests at heart, as they don't want to see you getting let down, but there are other people out there who will just hate your guts for no good reason. I knew that

the more my public profile grew, the more critics there would be. But I also knew that I had many more people cheering me on and wanting me to succeed than criticising me. These are the people you need to focus on.

8. There are some things in life that I know I will never be able to improve. That might sound pessimistic, but the reality is, when you're fifty-four years old, it is highly unlikely you're going to run as fast as you did when you were twenty-four. Of course, there are many examples all over the world of endurance runners who, at the age of fifty-four or even sixty-four, will beat people in their twenties and thirties. However, while you may still have great endurance at such an age, you will almost certainly lose your speed. There are some things that you just have to accept in life, and slowing down is one of them. I may slow down physically, but mentally I feel as strong as I've ever been. I acknowledge that my knees aren't in as good a condition as they once were, but I don't focus on that – and nor should you. If you focus on what is not going right and all the things that you have lost, then it can lead you down a very dark path and affect your mental

health. It is far better to focus on the things you *can* do and the areas you can improve in your life.

9. It's always important to recognise the people who have got you to where you are. This is true on both a small and a large scale – whether that be letting someone have right of way on the road, or the approach you take if you suddenly develop a big public profile. When people used to send me messages or comments about something I'd done, I would always acknowledge them personally, but, as my public profile has grown, it's a lot harder to do so. When you're followed on Instagram by tens of thousands of people, it's impossible to message everybody back. It's not that I'm not grateful for people's comments, it's just that I simply wouldn't have any time in the day to concentrate on my family or my business. When we were thinking of things to do in my theatre shows, we decided that, rather than leave the building and not acknowledge anyone, it was important for me to get out and meet as many people as I could. This is one of the things that Tom Cruise does brilliantly, which I saw with my own eyes as his bodyguard years ago. He was fantastic and it has paid off in spades with how the public sees him.

10. Never give up! I know many people look at me and think that I'm some sort of 'super soldier'. I will concede that I've been very fortunate to get where I did as a WO1 in the most elite regiment in the world, and that many people helped me along the way. But none of it would have happened if I had been a quitter. That is what you will understand as you read the following pages. I grew up the hard way, and that means not knowing when to stop, give in or surrender. That has never been in my DNA.

CHAPTER 1

The Man with the Hat

I'm running. I'm running fast, fast for a 9-year-old anyway. I was always an athletic, skinny sort of kid. I was into sports and, like most kids back in those days, spent most of my time out in the streets, doing physical stuff like running, biking, playing football. Me and my mates were active, especially compared to kids today, and we knew our area like the backs of our hands. That 9-year-old me, running like the clappers through the streets, knew where he was heading. But he was also starting to shit it, because the big fucker chasing after him was catching up. I was hanging onto his hat as hard as I could, thinking to myself, *Don't drop it, that would be giving in.* Instead, I was dodging, ducking, weaving, the sound of my little feet slapping hard against the ground below while his heavy thuds got closer and closer. My heart was ready to burst out of my chest as I ran, looking back to see if he was still there. He was. Fucker. *Why doesn't he give up?*

When I was nine, I ran with a gang of mostly older kids from the same bit of Walsall as me. It was rough and we were poor. I'd get up in the morning, no breakfast, but, if I could find a half-drunk cup of cold tea left over from the night before, then I'd knock it back before trying to find something to wear. In our house, it was first up, best dressed. So, I grabbed whatever clothes I could before my brothers, from shoes to socks to T-shirts, and off I'd go for the day's adventures.

Mum and Dad were both Walsall born and bred. She was Catholic, one of sixteen kids, and he was Protestant. They married when they were in their early twenties and had my eldest sister a year later. He was massive and she was tiny and they adored each other, but they couldn't have been more different. Mum was a saint – the sort of woman who never had a bad word to say about anybody. Even if I'd been beaten up or beaten someone else up, she only ever wanted to put things right and make sure that everyone was okay. If I got in trouble, my mum stood by me. No questions asked. We used to call her Ollie after the character Olive from *On the Buses*, because she'd laugh at anything and seemed to spend most of her time giggling at something funny.

Dad was the boss of the house and we all knew it. If Mum bathed us, the bath would be full of suds and she'd let us sit in it and play, but if Dad did it, then we'd get dropped into the scalding water and scrubbed like dishes to get it done as fast as he could. Dad always dominated while he was around, while Mum was the loving, caring one.

They were proud people, too, and would rather have died than ever sign on or claim benefits. Even though we'd probably have been better off if they did. Instead, they worked every hour God sent to make sure that we always had money coming in, even if it never felt like there was much of it around.

If my school shoes had a hole in them, I wouldn't get new ones. Instead, Mum would cut a bit of cardboard and stuff it inside to stop the rain coming in. That kind of thing meant that I learned to be resourceful. Once, I noticed that the lost-property box at school had a new pair of pumps in it, so I staked out the gym before school, lying on the grass opposite, until I saw the PE teacher go outside. With his back turned, I legged it in, nicked the pumps out of the box and ran for it. I wore those shoes every day until the next hole appeared.

My brothers and sisters were the same. I was the middle child of five, all squeezed into a three-bedroom terraced house in a quiet cul-de-sac in Walsall. Mum and Dad had their room, the three boys another and the two girls shared the box room at the top of the stairs. My elder sister, Beverly, ran the roost among the kids and was happy to give any of us a slap if we ever stepped out of line. My elder brother, William, was a rocker and I was a skin, so we had many a good brawl or two between us. My little sister, Emma, was a nosy, curious kid, and my little brother, Andrew, though we called him Totty, knew best where the fridge was, because he was bigger than the rest of us put together. We were a family of fighters and we fought a lot, both with each other and against everyone else, going at it like cats and dogs inside our own house¿¿, while making sure nobody outside the house ever picked on us either. We'd go to battle for each other first.

We had so much freedom to learn back then; space and time to go out and experience the lessons of life for ourselves. Even if we thought something might be wrong, we'd still do it. Just to try it, learn from it. We had the freedom to sample life and make mistakes. That's how we learned, that's how we grew. Outside the house, the little gang I ran with included kids of all ages from the local streets that made up our estate. There was me, Facey, Steve Scott and Brian Abnet, who were the regulars, and then there were other lads who'd come and go. There were kids in my area who got stabbed or shot – it was that kind of neighbourhood.

We all went to different schools but we hung around together, doing the same things, whether it was scrumping apples or stealing cars, or even breaking into places to nick things we needed. We used to break into sports shops and steal the football boots from the racks because we could never afford any.

Facey was probably the leader of the gang if there was one. He was an only child, a black kid who lived down the road from us with his nan. We met at Marlow Street junior school when we were five and, even though we went on to different secondary schools, we always stayed mates. Together we'd fight anyone and everyone and we never once fell out with each other, which was weird because I used to say that Facey could fall out with his own shadow.

Days would always start at the same rendezvous point at the end of our street, near the little park, where we'd wait until the last kid had turned up and then we'd be off. No real plan or direction, just out all day, creating havoc. Not deliberately trying to cause trouble but always sort of finding it. I'd gone rogue even by then. My father was an influential man in my life in many ways but, by nine, he'd lost control of me. The main thing we were all into that year was nicking trilby hats. It was 1974 and all the men wore them in the street when they went out. For us, nicking them was like a craze, the 'in' thing to do. I've no idea why. Maybe it was their association with American gangsters like Al Capone. Maybe it was just the sheer fucking audacity of nicking a hat off a grown man's head.

Five minutes before I found myself running for my life from the man with the hat, I'd seen him walking along the street – a tall old boy, seventy-odd, walking hand in hand with his beautiful wife, having a nice Sunday stroll down the high street, minding his business. There was nothing to mark him out except for what he was wearing on his head – a new grey trilby hat with a black silk band. It was like catnip to me back then, a prize too good to ignore. We may have been nine years old but we already operated like a pretty well-oiled machine. I was the fastest, so I ran point. Then the plan was clear: my two accomplices spread out to perform a classic pincer movement.

I wonder now if the British mentality that marks out soldiers starts young in us. I always think that British forces are at their best when there's a problem to solve. The British way is to ask first, 'Do we really need to go full-on here?' If there's a way to cut around the enemy, cut off the circulation of support, avoid direct confrontation, then we'll do that instead of taking the enemy head-on. Being clever has always served us better.

Later in my life I would see evidence of this in action from Iraq to Afghanistan. Maybe it's because we don't have the equipment of some of the larger armies, maybe it's because we don't have the money or we don't have the numbers, but whatever the reason, we've learned over many, many years to think first.

My instructions were clear: Facey and Brian would each come in from the flanks and create a distraction, because I needed the old boy with the hat to be thinking about anything and everything except what was going on behind him. They fanned out just like I told them, taking up wide positions until I gave the signal, and then *boom!*, they pincered in, knocking the old boy's concentration for just long enough to give me time to run up his back and grab the trilby off his head. I put it under my arm and ran. I figured, as this bloke was seventy years old and I was nine, he wasn't gonna chase me far. But he surprised me. Often people who we'd pinch stuff from back then would chase for a bit, have-a-go heroes with a rage in their veins, but they always gave up, ran out of puff, because they knew they weren't gonna outrun me.

But the old fella chasing me that day was made from different stuff altogether; this bloke was like Linford fucking

Christie. He even had a big old Crombie coat on and, trust me, running in a Crombie is like running in body armour. My mates by now had made a break for it, peeled off to the sides, never to be seen again, thank you very much. But the big man kept his focus on me, and on his hat. I pegged it along the Walsall high street, weaving in and out of the people, while he kept up with me every step, and I started thinking to myself that maybe this time I was in for a proper hiding.

When I was nine, the area that I roamed around in Walsall seemed huge. It was constantly wet and dirty and there always seemed to be a fog hanging over the place, but I would go miles from my house exploring new places. Me and the gang would roam the areas round by the pub where my dad drank, the railway lines where we used to lob stones at the trains, the arboretum where we stole apples and the canal where we sat around eating them.

The political situation in the country was shit; unemployment was rising, factories and businesses were closing, the 'Troubles' were raging in Northern Ireland and the miners were striking up and down the country. Walsall probably felt as grim as anywhere else to the grown-ups, but to me and my mates it felt like a big adventure playground. A derelict factory looks like a sign of urban decay to an adult, but to us it was just a place where we could jump out of the windows. Walsall was one big fun park that I had the keys to, and I knew every twist and every turn.

I kept running, turned left down a side street, right up the alley, left again, through the town, hoping to lose this giant on my tail, but nothing seemed to shake him. I must have been looking back too much because that's when I fucked up, turning right into a cul-de-sac. Suddenly cornered by a glass shopfront ahead of me, fences to either side and with no other road out, I was screwed. Instinctively, I turned around to face the danger. The fight-or-flight instinct was strong in me even then.

I bounced up into my fighting stance – legs spread, fists raised, up on my toes. I may have been nine years old, 5 stone 7 wet, and facing a fully grown man twice my size, but now that I took a proper look at him, broad and tall, built like a fucking tank, I was still gonna fight him. His hands spread out either side of him like Kenny Everett, making sure I couldn't get around him, and I said to myself, *Right, let's get on with it.* I was going to fight this man for his hat.

I've always had that attitude. Even years later, with SAS selection, I can remember standing on the square looking around again and thinking, *I am the skinniest, smallest guy here*, and I should feel intimidated, but no, I've always felt that the bigger they are, the harder they fall. Strength is in the heart and in the mind. That kid in Walsall was the same person as the man who turned up for SAS selection years later. I've always liked to see the big lads drop and I've always known that I had to do better than them. Right then, on that street in Walsall, that meant the scary-looking fucker standing in front of me. I wanted that hat and I wanted it more than he did.

He stood over me, like he was studying me all of a sudden, checking my feet first then working up, tilting his head to get a good look at my fists, nodding to himself until eventually he smiled.

'Keep the fucking hat,' he said.

I wasn't falling for it, whatever sneaky tactic this was. I kept my fists up, still bouncing on the balls of my feet. Damn right I was keeping the hat and I didn't need his permission.

'I'll tell you something,' he said, 'you little shit. There's something about you.' He thought for a second, then he said, 'So you keep that hat, but on one condition. You know the booths boxing club at the Digbeth pub?'

I did, of course. I knew exactly where it was. It wasn't far from my house.

'You come along there Monday night.'

It started to dawn on me that maybe he might not belt me after all. That was what went through my head. Fuck the boxing. Fuck the hat. I still had my fists up for a fight, but now this massive lump was going to let me off? That was new. Every time I'd been in trouble before it came with a belt of one form or another.

I was a kid who skived off school all the time, hung around my mates' houses or down in the town, down the canal, anywhere I could cause a bit of trouble. Sometimes I'd go down to the arboretum, scrumping apples, and not come home until ten or eleven at night. I was the kid who got hammered on whisky for the first time when I was nine. Abbo's dad was a drummer in a band and, when he went out, we'd often neck whatever bottle he'd left lying around. I remember stumbling in the door more than once, puking up all over the hall and getting a good hiding off my mum. She might have been the caring one, but she could put us in our place if we ever stepped out of line.

I was a kid who took a hiding often. Kids were slapped all the time back then and there was a real fear of the words 'Wait until your dad gets home' that we all lived with. If my dad was mad at you, then you could expect to receive the red hand of Ulster on the back of your legs, or worse. Sometimes he'd take the belt to me. He couldn't control me but, my God, could he hand out the punishment when it was required.

The punishments would settle me down for a little bit each time, but pretty soon I'd be back at it. I always knew I could get away with being naughty for most of the time. I figured that if I got caught one in ten, then I had nine other times I got away with it, and I could live with those odds. But now, standing in front of this big man, I realised that this was different. I picked up the hat and I threw it at him. He bent down to pick it up and, seeing he was distracted for a second, I took my chance and ran. This time I didn't look back.

—

Five o'clock in February; it was already pitch-black when I arrived at the Digbeth pub. Frost was in the air and I could see my own breath. There were no lights, nothing to guide me as I wandered into the darkness of the car park and around the back of the pub. I was shitting it. Proper terrified. I didn't know the first thing about how to box and I suspected that this might be a trap, thought up by the old boy to lure me to a spot where he could give me the kicking I deserved for nicking his hat.

It was the days of Muhammad Ali and George Foreman, the Rumble in the Jungle and all the big names, when the big heavyweights ruled the world. My dad loved boxing. He never boxed himself but he liked a fight. He was a big boy with a bit of a reputation locally, a 'name' I guess you could say, so people used to come down to his local pub, the Horse and Jockey, just to have a fight with him. My mum wouldn't even set foot in that pub. But my dad loved watching the big boxing matches on the TV too, and so me and my brothers would all gather round to watch them with him. That's all I knew about boxing.

Now there I was, standing in front of a massive, heavy wooden door more than twice my size, behind which there could be anything waiting for me. Still, I knocked on it and took a deep breath as it groaned open. Standing behind it was the big old boy himself, looming over me like Lurch from *The Addams Family.* He looked me up and down, making me take another shit in my pants, before he boomed down at me, 'Come in.' I think parents these days worry more about their kids and what they're doing than ours did. I wonder now if any parent I know would ever dream of letting their kid turn up alone at an unlicensed boxing club, behind a pub, because some bloke they'd met on the street had invited them along. Of course they wouldn't, but how different my life might have been had I not done exactly that.

I stood peering into the corridor behind him, knowing that this was my last chance to turn around and make a run for it. But I didn't run this time because I felt compelled; I felt I had to go inside. I followed him through another door, which opened up into a large room. It wasn't a gym, just a back room in a pub, with a bar along one wall and the tables pushed up to the side. There was a bit of a stage area where they'd laid out punchbags and skipping ropes and a bunch of kids, some my age, some older, were training, doing shadow boxing and push-ups.

Lurch pushed me towards them and told me to follow what they were doing, so I watched and fell in with their movements. Nobody said anything. I just started copying, throwing punches, side-stepping, bouncing on my toes; I felt surprisingly comfortable. I actually liked the feeling as the blood started pumping and my nerves started to settle. I could see I was the smallest kid in the room but I was doing exactly what the bigger kids were doing and that felt good. I didn't know it yet, but that would become a familiar feeling throughout much of my life.

After an hour of training with the other kids, Lurch came back and called me over to him again. 'Come with me,' he said, so I followed him once more, this time to a smaller room, where he put me up against the wall and I started to feel the nerves rise up again.

Fuck, I thought, *this is payback.*

'Put your fists up,' he said, showing me what he meant.

I looked to the sides of him to see if there was a way to slip past, but he was blocking the exits so I decided to just do as I was told.

'Now, put your right fist to your right ear,' he said.

Again, I did as I was told, keeping one eye on any potential exit. Quick as a flash, he wrapped a strip of gaffer tape around my head, strapping my fist to my ear so I couldn't move it.

Shit, I thought, *now he's really got me.*

But he didn't hit me. Instead, he started to teach me how to box. How to *really* box. How to hold up my head. How to duck. How to weave. How to move from side to side. He spent an hour with me, just me, instructing me, passing on what he knew. When I got things wrong, he made me do them again until I got them right, and when I got them right, he praised me. I worked hard and it felt good.

It was the first time that someone outside of our family had taken time to show me they cared about me. The first time someone ever thought to give me some purposeful direction. Yeah, my dad was always there and I loved him, but he was always working, always out making money to keep food on the table, or down the pub drinking. So this felt like something entirely different. This was a man I hardly knew and he was giving up his time to teach me, a kid who'd tried to rob him, how to do something that I was enjoying. This was the start of something that would change my whole life.

—

I started going to the boxing club behind the Digbeth pretty regularly, most Monday and Wednesday nights. I wasn't like a clockwork regular, but I'd go as many nights as I could. The club was classed as a booths boxing club, which meant it wasn't accredited to the Amateur Boxing Association. The booths clubs were not unlike the kind of unlicensed boxing often associated with the Romany community, and the fights that they'd train for were bare-knuckle and illegal and happened on what were called the 'gambling nights'. In their heyday, the boxing booths would travel around the country on the fairground circuit and the boxers would actually fight for proper championships at regional and national level. Proper hardmen like Harry 'Kid' Furness and Lenny McLean were known throughout the country. Those guys were legends in their own lifetimes but, by the 1970s, the booths were starting to disappear, and there were only a few clubs, like the Digbeth, that were hanging on, carrying on that tradition, bringing on the new lads from poorer backgrounds.

Not that any of that bothered me. I just enjoyed being there. I'd turn up and train hard, doing all the things you'd do in a normal boxing club: shadow boxing, bag work and my favourite thing of all – sparring. Even at nine, I was getting the hang of sparring and actually I was already boxing pretty well. With a bit more coaching from the old fella, I even started holding my own with some of the older lads and reaching a good little standard. Then, one evening, he came to me and told me I was up next.

I was put up to spar with the club hero, a lad who was a year or two older than me, called Gary Bullock. Gary had been training at the club a couple of years already, but I wasn't scared of him, even if I probably should have been. I actually couldn't wait to go toe to toe with him and see how I did.

When our turn came, I got up in the middle of the circle, all the other lads gathered around us, cheering, shouting and screaming. We started circling around, throwing a few testers at first, sizing each other up, until a couple of seconds later, *wham!* – I hit him. I hit him square in the jaw and I put him down, flat out on his arse. For a second everything went quiet; I stood looking at Gary lying on the ground in front of me. What had I just done? Suddenly, everyone went wild.

Knocking Gary down meant I started to grow a reputation at the club and developed a bit of a name for myself. Winning had also spurred me on to train even harder and try to improve my standard, so I started going along to the club more regularly, working harder and watching the older fighters to see how they moved and hit. Then the old guy took me aside one night and said he wanted to talk to me.

'You've obviously got an aptitude for this,' he said, 'but I can't take you any further.' I was stunned. I was working hard and now was he throwing me out? 'It'd be good for you to fight in competitions,' he said, 'if that's what you want to do?'

Now I was confused. I did. I wanted that so badly. Sparring was my favourite aspect of the training, and I loved it when I got to stand up and go toe to toe with another kid. I'd seen the big boxing fights on the TV and I'd dreamed about being in the ring myself.

'You can stay here if you like.' He said it like he already knew the answer. 'But if you want to go on and win competitions, win trophies, then we have to find you a proper club.'

Slowly it dawned on me: he wasn't throwing me out; he was actually showing me that he believed in me, believed that I had potential.

—

The Bloxwich Amateur Boxing Club met in an old school gym in Walsall. The old fella took me up there himself and introduced me to Bill, Eddie, Henry and Freddie, who ran the place. They promised that if I started training there, then I could work towards having my first proper fight. I wanted to get in the ring. I wanted to test myself and I wanted to show everyone how good I was. They advised me that moving to the Bloxwich club was the best way to do exactly that.

Bloxwich was a proper boxing club. Everything about it was more organised, more formal than I'd experienced before and, even though it was classed as an amateur club, to me it seemed completely professional. There were no more tables pushed up against the wall and it was fully equipped with punchbags, ropes, a weights room and even a proper ring. The club was regulated by the national Amateur Boxing Association and the lads training there ranged from youngsters, 11-year-olds like me, right up to proper men – men training for real competitive fights. We were all assigned to an accredited trainer and every week we did skill-set training, weight training and boxing.

There were a few kids at the club I knew from the streets round where I lived, but it was the first time I'd hung out with other kids who weren't from my area – kids from other schools and other towns in the West Midlands. We trained and sparred together twice a week and, for the first time in my life, I felt like I was part of something bigger than just my gang. I was now part of a real club.

About eight or ten of the lads were deemed to be of a standard good enough to have a competitive fight in the ring and, because of the training I'd already done at the booths boxing, it didn't take long before they put me in that group.

The night of my first fight came around fast and we all travelled together in a minibus to the Walsall town hall. As we arrived, I could see the room was jam-packed, the whole of the West Brom and Walsall football teams were there, as well as an ITV News crew. My mate Mark and I had been chosen to fight for Bloxwich in the lightweight categories that opened the bill, so as soon as we changed, we were taken downstairs to be weighed in. It was nerve-racking; so many people, so much noise, grown men smoking cigars, sizing us up and placing bets on us.

While they were weighing us, one of the coaches from the club came over with a long face. He had bad news. There weren't enough people in our weight category to fight. I was gutted. It meant my fight was being called off. I was nearly in tears because I was absolutely ready now, mentally and physically, to get in that ring and fight. Mark was the same.

'There is an option,' he said. Mark and I were both the same standard, same age, same weight; neither of us had fought before but we were both ready to.

'You can fight each other,' he said, 'but Billy will have to fight for West Brom.'

This was really fucking annoying. Why had they chosen me to fight for the other club and not Mark? I thought they must actually think Mark's better than me because otherwise they wouldn't have done that, which made me determined to prove them wrong.

Minutes later, Mark and I climbed into the ring, with hundreds of grown men around us shouting and cheering. We were just two little lads, hardly bigger than the ropes, but for three one-and-a-half-minute rounds, we went at it like Ali and Foreman. We'd sparred before but now grown men were snarling and spitting orders from their seats, coaches were urging us on and the bell was ringing out over the noise, while we went at it like a couple of wild dogs, toe to toe, ready to kill each other. It was totally different to any fight I'd ever had before and I went at it fucking hard. I went at my mate like he was my worst enemy. And he did the same. I knew one thing – I wasn't going to lose – and he knew one thing: neither was he.

But it turned out he was wrong, because he did lose and I won. In front of the West Brom football team and all the bigwigs in their black ties, I won. People came running into the ring, there were hugs from people I'd never met, others lifted me up, and everyone was slapping me on the back and telling me how well I'd fought. As I came out of the ring, the butcher who owned a shop down the road from us came up and stood over me.

'Great little fight, son. We just won some good money on you.' He gave me a pat on the back and then handed me a voucher for free meat from his shop.

That night I went home and I gave the voucher to my mum. She couldn't believe it – free meat was a really big deal in our house. I was so proud because I'd never won anything before in my life and now I'd fought, won a competitive fight and had something to contribute to the family. My dad never said anything about it, which I couldn't really understand. He never came to see me fight either, forever saying he was working. I decided he just wasn't interested in me and what I was doing, so I resolved to work harder, fight harder, to prove him wrong. I wanted him to be proud of me, and somehow I'd show him.

CHAPTER 2

I Don't Think You'll Ever Go Anywhere, Billingham

Around the same time as I started boxing, my elder brother came home one night talking about joining the cadets. My brothers and I fought a lot, but we also got each other into things. Sometimes they'd come boxing or play football with me and my mates, or I'd go with theirs. When I heard my brother talking about the cadets, I was curious. I thought, *Why not go along and see what that's all about?*

Walsall Marine Cadets comprised thirty marines, thirty sea cadets and fifteen wrens. Every Tuesday and Thursday night, on an open piece of land near the centre of Walsall, they assembled to parade on a square of dirt next to some old Portakabins, in front of a big red-brick Victorian building that looked a lot like the factories my mum and dad worked in.

Mac was the man in charge; a huge, solid, square-shaped guy who stood around 6ft. He was hard as fuck; nobody messed with Mac Gaunt. He would tell us how it was and, even though we were all very bad kids, cheeky kids, it was clear that if Mac told you something, you did it. If you did ever step out of line, then you'd get a clip round the ear for it. To begin with, when I first went along and I was being shouted at, I thought it was all a bit of a giggle, but soon I realised that Mac wasn't joking about. Mac talked to me like a man, like an adult, like a soldier. He was very serious and that meant that I wanted to take it seriously.

The recruiting process lasted for six weeks, during which time I learned how to read a map, how to strip and assemble a gun, as well as what a gun could do. I learned basic soldiering skills of camouflage and concealment, how to carry out observations by hiding, how to report and how to do signals. I was drilled in communication skills, first-aid skills and all the things that would be required of a soldier, just like in the army. Then I took a test at the end, knowing that if I passed it, I'd be given the uniform.

One thing I'd never done before, though, was pass a test. In fact, I don't think I'd ever even turned up for a test during the whole time I was at school. When the day of the cadet test came around, however, I turned up for it early and applied myself like never before. When Mac Gaunt gave me my cadet uniform later that day, it was the happiest day of my short life, even happier than winning my first boxing fight.

Shortly after that, Mac took us on our first camp up in the woods around Cannock Chase, where we lived in a bunch of old log cabins, like we were in a real army camp, for two weeks. Every day, we went out in the fields, digging trenches, playing soldiers. Mac started sending me away on extra courses too, so that, during weekends and school holidays, I was doing navigation courses, rope technique courses, even a sailing course once.

I loved it. I really fucking loved it, because I'd never done anything like it before. I knew in my heart from the start that I wanted to be there. As with boxing, the cadets brought me into contact with a whole new bunch of kids, from eleven to eighteen, some of whom I knew, some I'd heard of, but most I didn't really know at all. It was another whole new world to me; structured, ordered and disciplined from the word go. I liked the camaraderie of being in that little gang, doing something good and hanging out with kids I didn't have to fight. Now we were all fighting for the same cause, doing the same things together, like a real team that I was enjoying being a part of. I'm still friends today with many of the lads I met in the cadets. When I got married, ten of them came to the USA for the wedding.

In the cadets, I learned how to run with weight on my back like they do in the army, which meant I was doing a different type of endurance fitness to my training at the boxing club. The combination of those two really set me up well for the army in later life. The skill sets and discipline that I learned in the cadets together with the fitness and courage I gained in the boxing ring undoubtedly put me at an advantage when I got to the military.

Mac pushed me as far as possible to learn new things. He would show me how to do something, ask me to do it, and then make me do it again and again until I could do it to the best of my ability. It felt good for a kid who hadn't done well at school to find something I could be good at. I thought I was good at football (but I really wasn't) and I was doing well at the boxing, but cadets felt different; cadets felt like something bigger, something I could really excel at.

When he wasn't running the cadets, Mac worked for the council, but he took time off for his cadets. If any of us ever got into trouble and needed him, Mac was there for us. He put time into us, which made us feel like we had to give him that time back in return. Mac taught me cadet skills, but he also took an interest in guiding me through the cadet structure, first as a private, then as a corporal, and finally as a sergeant. He helped me to understand what it meant to think like a soldier.

People often ask me what it means to be a good soldier and I tell them what I learned from Mac – a good soldier is someone who's loyal, has good discipline and good skill sets, but a good soldier is also someone who naturally becomes a leader. You can look at a good soldier and see someone who isn't scared of making a decision and know that the decisions they make are generally going to be sound. A good soldier should be someone you'd be happy to follow, work with and trust. Soldiering isn't about killing; it's about character. Watching and growing as a person is the most important experience a soldier can gain and I started learning that in the cadets from Mac Gaunt.

—

As I got older and became a teenager, I started to get into my fitness and, even though I probably wasn't eating properly by today's standards of what an athlete should and shouldn't consume, I did try to make sure that I was resting or not eating too many chips in the build-up to a fight.

Around that same time, sniffing glue became a big thing around Walsall. Young lads and young girls, too, were filling plastic bags with glue and then inhaling it until they hallucinated and eventually passed out. As far as I could tell it was a fucking stupid thing to be doing, but, still, a lot of my mates and their friends took it up, got addicted to it and messed themselves up.

One afternoon, when I was training for a fight, running back along the canal, I looked across the football field and saw something odd. It looked like a dead body in the middle of the field, so I ran over towards it. As I got closer, I could see it was my mate, Ade, a lad from the boxing club, lying out flat on his back with bags of glue all around him. Ade was a great guy, with real potential, but now suddenly, although he was only thirteen, he looked twice that. He went downhill fast; he stopped turning up for training, his fitness dropped and he ruined his life. Glue was horrendous, yet still people kept doing it.

But not me. I think because I was so into boxing and staying fit, getting ready for fights, I just never wanted to do it, so I never even tried it. It's something I'm still grateful for today.

I wasn't into drugs either, but I was still a part of the whole music and gang scene. My gang were rude boys. Skinheads.

We shaved our heads, turned up our jeans, wore braces and red laces in our Dr Martens boots. There were other gangs around, which we didn't get on with – rockers, mods and the punks. It was just like the movie *Quadrophenia.* The gangs would often get into fights with each other, and, while it was all fists and boots at first, before long I started to see knuckle-dusters and, eventually, knives and cut-throat razors making an appearance. We all thought we were ballsy but we were actually just stupid. We'd go and fight with other gangs all the time, Facey always the first guy straight in there, literally fighting anybody, no matter what age, size, reputation. Facey would go for it and then I'd follow in behind.

I started to enjoy fighting more and more, getting off on the adrenaline and the fear until I felt like it was almost the everyday, normal thing to do. Boxing had helped me to be a better fighter, teaching me how to read people's faces. I always say boxing is the poor man's game of chess because you learn how to watch someone, size them up, how to prompt and how to counter. By the time I was fourteen, I could tell by somebody's face when things were going to turn nasty, when they were going to make a move. As I got older and better at boxing, I got better at street fighting too, until I was conditioned for a fight any time and place, confident that I knew what I was doing, whatever the situation.

I think fighting helped me prepare mentally for the army, too. Fighting and boxing gave me grit, balls, the desire to go for a challenge and see it through to the end. It also improved my fitness, because I was training really hard most of the time.

But as the fighting became more violent, more knives appeared and things began to change. One night, me and my mate Patrick went along to a skinhead disco where we knew a load of rival skinhead gangs were going to be coming in from all over Birmingham, Walsall and Wolverhampton.

The funny thing was, when we got in there, we actually kind of liked the music and started to have a bit of a dance. I remember saying to Patrick, 'This music isn't too bad, you know?' – just as the music went off and the lights went dark.

Just then, in the dark, I felt a lovely warm feeling flush all over my face. A second later, the generator kicked in, the lights came back on and I could see a lad standing in front of me, dressed in his skins gear, holding a cut-throat razor by his side, blood running down his arm, dripping onto the wooden floor. I turned round to see Patrick's face had been slashed open from ear to ear, half his nose was hanging off and the lovely warm feeling I'd felt was his blood all over me.

Panic gripped the venue. There were screams and shouts as everyone scurried off in different directions as the police arrived and Patrick was taken to hospital to have his face stitched back together. As per usual, with all the violence that permeated our lives, nobody ever got charged for it.

—

The following summer, I turned fifteen and went back to boxing again. Boxing was always like that for me, with periods where I was really into it and periods where I could take it or leave it. That summer I built up to a couple of fights, won both of them and then dropped out of it again, drawn back into the gang, getting in trouble, getting into street fights, losing direction.

School wasn't much better. Boxing had trained me to be a fighter, but I was frustrated that I was only getting one competitive fight a month, so I started looking elsewhere for fights outside of the ring. I wanted more and more and school was a good place to find them.

Once or twice a week, I'd disappear around the back of the gym for a fight with somebody. I developed a really bad reputation, but didn't care, because all I wanted was to be someone who didn't take any shit from anyone. At fourteen, I was fighting 15- and 16-year-olds; by fifteen, I was fighting 18-year-olds. I'd fight bunches of kids, whole groups on my own, anyone I could find who thought they were harder than me – I'd fight them all and even came off a silver medalist some of the time!

I got into a fight with two bullies in the car park of the British Legion club, a bar where my dad used to drink. I knew they had a real reputation as fighters, so I turned up specifically to fight with them; it was prearranged and clear what we were all there for. I took them both on together and beat the crap out of them. Some of the older guys from the bar, mates of my dad's, came outside to watch, egging us on and cheering. When my dad got to the pub that night, I heard that his mates were describing the fight to him like I was a hero.

Every day felt like I was trying to win a title, like I always felt I had something to prove to myself, to the other kids, to my dad, to everyone. I wanted to be seen as a leader, to be known as good at something. I wish I'd been better at football but, instead, I was good at fighting.

I even won my first wife in a fight. After what had happened to Patrick, I was trying to avoid discos and clubs, but a mate of mine, Stuart, who'd applied to join the navy, told me there was a rock-and-roll disco that some of the local girls around town were going to, so we decided to give it a go.

The skinhead scene was starting to fade away by then and the Teddy boys were becoming a force; with all their fancy suits and their big hair, they fancied themselves as ladies' men. There was one girl I noticed on the dance floor right away and took an instant shine to. Her name was Julie and I fancied her, but, when I showed her some interest, a big Teddy boy lad, a couple of years older than me, pulled me to one side and told me that he was already going out with her. He tried to front me down, so I said to him, 'Okay, I'll fight you for her.'

Of course, I put him on his arse, so Julie became my first girlfriend and, four years later, my wife.

By fifteen, I'd fought everyone in the school, so I couldn't see any reason to keep going. I certainly didn't have any interest in studying and all the teachers knew it. There was one teacher, Mr Noone, who was always on my case, trying to discipline me for fighting or playing truant, so one time he locked me in the economics room after school for detention.

'What are you going to do with your life, Billingham?' he asked me.

'I'm going to join the army,' I said.

'And what if they don't want you?'

'I'll join the navy then.'

'You're a cocksure little shit, aren't you, Billingham?'

'Or the RAF and, if they don't want me, I'll go in the merchant navy.'

I'll never forget what he said to me next: 'I don't think you'll ever go anywhere, Billingham, but that's up to you.' Then he left the room and locked the door behind him, so I climbed out of the window, jumped down into the bushes and went home.

When I was bored or playing truant, I'd hang out along the embankment that ran behind my house. All the gang would come and go, messing around and causing trouble. Sometimes we'd go down to the line, fill a bag with stones and bits of brick, and pelt the passing trains. Or, if there weren't trains, then we'd bombard our rival gang on the opposite embankment. More than once I'd been smashed in the face by a flying rock and ended up with stitches.

When I was fifteen, one of those scraps sparked two of the rival lads, twin brothers I knew quite well, to come running down their side of the embankment and chase us along our side of the tracks. They were older and bigger than us, so my gang turned and ran, but I decided instead to stand there and fight them. We went at it until one of the twins grabbed me by the ankle and I slipped. They took their chance and both grabbed hold of me, dragging me down the bank and, as I kicked and fought as hard as I could to get away, one of the twins jumped up on my back, pinned me down, and his brother stabbed me with a knife, just below my right kidney.

They ran, leaving me for dead as I lay bleeding on the grass bank. The pain was incredible, but I knew I had to move or I was going to die there, so I somehow found enough strength to crawl back up the hill and along the embankment, bleeding all over the place. Eventually I reached our back-garden gate and managed to crawl to the back door of our house.

My mum must have heard me, because she came outside and found me lying on the step, bleeding to death. She went into a bit of a panic, ripping my top off, screaming at me, trying to find where all the blood was coming from. She grabbed the nearest thing she could, a mohair sweater, and pressed it into the wound. The last thing I remember, before I passed out, was that it stung like hell. They rushed me to hospital in an ambulance. When I woke up, the doctors said it had been very close – if I'd lost any more blood, I'd have died.

—

When I got out of hospital and back on my feet, I decided that school was a waste of time, so I might as well focus on trying to find a job. The problem was that I was still only fifteen, so I had to find someone who'd pay me cash in hand and overlook that working at that age was illegal. For most of my childhood, my dad worked twelve-hour shifts in a foundry where they made big metal chains, and my mum worked down the road in another factory where they did electroplating. He'd mostly work the night shift from six at night until six in the morning, while she worked from six in the morning until six at night, which meant that most of the time, me and my siblings hardly saw either of them. They were both real grafters, always working because they had to be; we were frigging poor.

One thing my father taught me was the ethics of work. As far as my dad was concerned, you worked. It didn't matter if you had a broken leg, he didn't care – you could put a wheel on it and wheel yourself there. If I wasn't going to school, then there was no other option: I was going to go to work. The factory that took me on was also an electroplating business, set in two large brick buildings either side of an office, behind which was a skip with an old Hillman Imp wedged into it, which the boss's son used to sleep in sometimes when he was too pissed to make his way home. It was that kind of crazy place.

I'd had part-time jobs before, working on a milk round for a while and then on the market some weekends, but this was a proper job doing adult shifts. I was still a year below legal working age, so it was all off the books, but still I was taking home 80 quid a week, cash in hand, which was more than my mum made. For the first time in my life, I had money, and, even though I was giving most of it to my mum, I felt like I was behaving like a man.

The factory was also where I met Joe. Joe Taylor was an old boy in his sixties who worked in one of the workshops. I knew he'd been around for years because he knew my dad from way back. Joe's head was covered in scars but he never talked about how he'd got them. In fact, he didn't say a lot, except every now and again, when he'd sit me down and ask me what I was going to do when I was older. Of course, I didn't know, but I liked that someone there was showing an interest in me.

One morning, a big delivery truck rolled into the yard to pick up some plating and the driver climbed down from the cab and told me to get the forklift ready. But before I could go, he stopped me. 'Is that Joe Taylor?' he asked. There was something about the way he said it, a sort of respect in his voice that surprised me. 'When Joe was on duty, everyone slept better,' he said. 'We were all safe when Joe was on.' I didn't even know Joe had been in the army.

Joe showed me the ropes at the factory, explaining to me how electroplating worked, taking me through each of the huge metal vats filled with chemicals and acids for the electroplating. There were vats of acids and cyanide and caustic soda, which would strip anything you put in them. Whatever we were 'plating' at the time – from pots and pans to car parts or electric goods – we'd hang the items off big metal hooks and drop them into the vats, one after another, leaving them there for a few minutes before we moved them on to the next.

Back then, there was no health and safety of course. I was allowed to do everything the grown-ups did, even drive the forklift. One time I thought it would be funny to drive it full speed through the factory doors, totally forgetting that the top of the forklift was higher than the doorway, so I smashed into it, bringing a tsunami of rubble down all around me. I sat there, covered in dust, while the women in the factory went screaming and running in all directions. The manager came flying out of his office, took a look at what I'd done and said, 'You're gonna have to work overtime to pay for that.'

Mostly, I'd work the night shift, helping out on the crane that lifted the jig – a metal rack that carried all the items we were plating. The crane pulled the jig along and I dunked it in the first vat of acid for a couple of minutes and then helped it back out and along to the next vat for the same thing. Often the jig would get stuck, so whoever was on the crane had to reach in and pull it out manually.

One night while I was operating the crane, the jig had got stuck again, so I stepped down onto the side of the vat to grab hold and pull it out. I must have lost my balance, because the next thing I knew, I was falling, dropping feet first into the vat of caustic soda, immediately feeling the chemicals cutting through the cloth and a horrible burning sensation running up from my feet.

I knew that if I was going to survive, I needed to get out of that vat, but, to this day, I've no idea where the strength came from. Somehow, despite the immense pain, I pulled myself up and over the edge of the vat. I fell out the other side and there was Carl, one of the crane operators, to catch me. As I lost consciousness, I could feel him carrying me like a baby towards the water hose.

Carl had to act fast, so he threw me down on the concrete floor and started tearing my wellies and trousers off, running a hose over me. I was passing in and out of consciousness with the pain, but I can just remember the sight of my legs, the skin bubbling up full of fluid like two huge balloons, and Carl saying to me, 'We'll get you to hospital, mate, but you'll have to say you broke in – you can't say you were working here or we're all fucked.' I was only fifteen and I'd narrowly escaped death for the second time in my life. I didn't realise it then, but it was a sign of things to come. I did what I was told and stuck to the story: I'd broken in and fallen into the vat. It was my own fault; nobody else was to blame. The police never came to question me, so the factory owner gave me a nice bonus for keeping my underage working a secret and I went home to recover.

After the accident, Joe came to see me and said it was a sign that I couldn't keep working in the factory for ever. He said that the army was going to be my saviour and he made me promise that I'd apply to join as a young recruit at sixteen. The problem was the accident had left my legs badly scarred and covered in holes, so that when I went up to Wolverhampton for my medical, they took one look at me and sent me away. They told me to come back when I was seventeen and maybe I could try again as an adult soldier.

—

Boxing and the cadets were my safety net during a time when I didn't have anything else. I'd already begun to realise that I needed to get out of Walsall. I'm not blaming the people around me, but I knew in myself that if I didn't break the cycle I was in, then I was going to end up inside or worse.

I needed a new direction, needed to hang out with a new crowd, and I suspected that, out of boxing and the military, the army was the way to go.

Boxing was definitely an option, but the army felt more full-on. I knew that, with boxing, I'd get time to myself between fights, which is what frightened me. In between the training and the fights, I'd still have time and space around me. Even if I became more channelled, more disciplined, I would still be vulnerable to slipping back into the gang, and I didn't feel like I could trust myself to stay out of trouble during those periods.

On the other hand, the military meant that the space I was in was my whole space. The army meant 24/7 commitment and I knew that if I could get in, then it would keep me occupied, which deep down is what I knew I needed. I liked the discipline and the structure, and I liked being told what to do by people I respected.

That might seem at odds with the kid who never turned up to school, had trouble sitting through a maths lesson or an English class, but there was a big difference. In the cadets, I was enjoying what I was being taught. I could see a practical use for it, where I never could with maths. I could triangulate on a map, find a bearing and hike to that location with pinpoint accuracy, but call it trigonometry and ask me to find the hypotenuse and I wouldn't have a clue. The difference now was that I could see the relevance of learning, and somehow that gave me the ability to do it, to do things I couldn't do in the classroom. It's a lot easier to apply yourself to something you're good at, and I was good at the things they wanted me to learn in the military. Teachers at school often accused me of having a problem with authority, but that wasn't true, because in the cadets I respected the structure and the discipline.

By the time I turned sixteen, I still had no idea which area of the military I'd end up in. I knew some of the cadets older than me had gone on to join the marines or the guards or the engineers, but I still didn't really understand what the difference was between them.

When the Falklands War ended and some of the ex-cadets who'd served started coming back, Mac organised an evening at the cadets' hall for them to come and celebrate being home. Part of the evening involved all us young cadets sitting down and listening to the soldiers talking to us about their experiences. We sat and listened while they told us their stories one by one. My jaw dropped to the floor as I heard how they'd been in battles, had been shot and injured. To me they all seemed like real heroes, but one story in particular stood out.

Frank was in 3 PARA and I knew of him already because his two brothers had also graduated from the cadets into the military. When it was Frank's turn to speak, he told us the story of the assault on Mount Longdon outside Port Stanley in June 1982.

The Falklands had been a ten-week conflict between Britain and Argentina that raged through the spring of 1982. Like every kid in that room, I'd followed it on the news. When Argentine forces landed on British soil, our boys had gone over there to claim it back. The fighting lasted nearly three months until the Argentine surrender on 14 June, when the islands returned to British control. One of the last battles was on Mount Longdon (11–12 June), part of the final assault on Port Stanley, capital of the Falkland Islands.

Frank sat and told us how he and a small squad had made an assault on an area of high ground on Mount Longdon, when they came under enemy fire. As he described to us the realities of being at war – of being outnumbered, trying to sneak up on your enemy, how his friend (Corporal Brian Milne) had stepped on a mine and blown his leg clean off, which had then alerted the Argentines so the whole mountain had lit up with enemy fire aimed right at them as 3 PARA had to then quickly make their way up treacherous slopes to take on the Argentines – you could have heard a pin drop. Frank had himself been shot and I was terrified for him as he recounted his story.

He told us about the camaraderie of his mates around him, staying with him all night and protecting him until help came. How 3 PARA had eventually made a bayonet charge to drive off the Argentines and capture Longdon. It sounded like the most awful time of his life and yet I found myself feeling almost jealous of it. I kept thinking to myself, *That's how I would be, that's what I would do.*

I sat there listening, knowing now that I wanted to be challenged like that. It was weird, because I obviously didn't want to be shot, but I knew I wanted to be challenged. I wanted to be in a situation where I had to make big decisions, make a ballsy 50/50 and see if I could win it, see if I could get to the top of that mountain. As Frank described his story to me, I kept imagining myself in his situation and a voice inside me started telling me that I probably wouldn't have got shot; I would have got up there, I would have got to the top of that mound and taken out my enemy.

The other soldiers' stories were impressive too, but on hearing Frank's story I felt a deeper connection. There was something meaty about it. I knew then that was what I wanted. I knew that the regiment he fought for, the Parachute Regiment, and specifically 3 PARA, was going to be the one. I had the answer to where I wanted to enlist.

I went up to Frank afterwards and told him that I wanted to be a paratrooper like him. I could tell that he was proud to hear that and that he wanted to encourage me, but he was also brutally honest. He said to me, 'It's still a distant dream for you, mate, until you've passed the training.' Of course I'm biased, but I really believe paratrooper selection is the hardest infantry training in the military. There are loads of lads who dream of being a para but just never make it. Frank probably felt the same way, because he offered to help and encourage me as much as he could, but in the end he said to me, 'It's all down to you.' Funnily enough, it was the same thing my teacher Mr Noone had said.

I still had two problems to get over first. I'd been up to the army careers office in Wolverhampton again, just like they'd told me to. The burns on my legs had recovered enough for me to apply and they'd given me all the stuff to read to help me make my decision, but they'd also told me I was way too light to join the Parachute Regiment. I was still barely 8 stone and the careers officer had absolutely ruled out me joining the paras. The other problem I had was that I'd lied on my medical about my previous injuries, and, while my skin was clear, they had found a fractured femur on the X-ray from when I was a toddler.

I was one of those kids who climbed everything I could. My parents used to say that if they took their eyes off me for a second, I'd be up on top of the nearest table or wall. When I was eighteen months old, I climbed up onto a wall, fell off and broke my leg. Now I had to go to London to have it checked out by a specialist. While I was waiting for the appointment date, I was eating like a horse, trying to put on weight, and my mum even bought me a tub of protein weight-gain powder to supplement my meals. I totally stopped boxing training to reduce the calories I was burning, but still nothing stuck. I finally got the appointment to have my leg checked and, even though the doctor cleared me to begin infantry training, the careers officer (who was a guard, I later found out) was adamant that, at my weight, the Parachute Regiment would still never accept me.

So, when I got the letter to join the first round of infantry training in Sutton Coldfield, I decided I'd go and just see what happened. I had tests for maths, English and all the things I hadn't done at school, having never turned up for a lesson. Of course, my spelling was atrocious and my maths not much better, but I somehow scraped through to the physical fitness test. When we arrived at the fitness ring, the sergeant told us to sit in our regiments, infantry over there, signals over there, Parachute Regiment over here, so I kept my head down and quietly followed the other lads going to the para section. Nobody challenged me. The staff came along the line of boys and handed me a band, the same band as the other lads, which meant I could do the fitness test with the other para guys. That was all the chance I needed. I decided that I would give it everything, do the fastest time, more sit-ups than anyone else, more pull-ups too – in fact, they had to stop me doing pull-ups because I wanted to prove to these guys that, even though I may've been half the size of them and I was shit at spelling, I was still the fittest bloke there.

The next month, a letter arrived at the house in Walsall. Inside were my papers to join. I was to report to Aldershot to begin Parachute Regiment training on 23 October 1983. The night before I was due to leave for Aldershot, my old man called me downstairs, something that usually only happened if I was in trouble. We sat in the living room and, for the first time in my life, he gave me a father–son chat. My dad explained to me that if and when I got injured on the battlefield, it would be him and my mother who would have to look after me. As he talked, I could feel how focused he was on all the worst things that could happen to me and not the best. To me, all I could hear was his certainty that I was destined to fail.

I was angry. This was the man who never came to any of my football matches, never came to any of my boxing bouts, even when I fought in the ABA championships, and yet here he was telling me now that I wasn't good enough? That I was going to mess it up? That I was being inconsiderate because I was going to get shot or wounded and come back needing his help? It was horrible, sickening, and when I went to bed that night, I couldn't sleep a wink because I lay there wondering whether he was right. Maybe I should give it up. Maybe it was wrong of me to do this to my mum after all the shit I'd already put her through over the years. Maybe I would get killed or injured. But something in me changed in the small hours of that night. I found something inside of me that said, 'No. I am good enough. No. I'm not going to fail. No. I'm not going to come back here needing anyone's help.'

I'd got to where I had the hard way. There was no doubt that I'd failed at school. I'd nearly died in a knife fight, almost lost my legs in an industrial accident and could have been beaten to death in any number of fights over the years, but I hadn't. I had made it this far against all the odds and I was going to keep going. I had further to go and I wasn't going to give up now. 'Always a little further' was already becoming my mantra.

When I woke up the next morning, I knew I was going to Aldershot and I was going to join the Parachute Regiment training, pass it and come back here a success. I was going to prove my dad wrong and show everyone what I could make of myself. I took my bag, kissed my mum goodbye and left for Walsall train station to begin what I knew would be the next chapter in my life.

CHAPTER 3

How Many of You Piss the Bed?

Even on the train to Aldershot that morning I was still thinking to myself, *Have I done the right thing?* Maybe my father was right and I wasn't up to this. The doubts were creeping in and taking hold to the point where I was almost ready to say, 'Fuck it', jump off the train and go home.

But as we got closer, I noticed a few lads who looked about my age and kind of fit getting on the train and I thought, *They're going to the same place as me.* One or two of them let on to me and I could tell from their accents that they were from the Midlands like me. One of them came up to me and asked if I was going to Aldershot, too. It suddenly hit me that that was exactly what I was doing. I nodded. *Damn right I was going to Aldershot.*

We all got off the train together and I could see right away that there were more people from all over the country who looked like us, waiting outside the station. I took a look around, sizing everyone up, and I could tell that we were all going to the same place. A convoy of minibuses pulled up next to us. The door to the first one burst open and a monster of a man jumped out, full uniform, beret on, and asked us, 'Recruit training?' Nobody spoke, we just nodded. 'Right, you on that minibus, you on that one.' He directed every one of us until we were all loaded up, and off we went to learn how to be soldiers.

Aldershot is an old garrison town and the depot is right at the heart of it. Back in 1983, they hadn't started to put up the fences around the accommodation that now make it seem more like a fortress, so out of the window I could see right across the parade ground to the open camps. I saw military types milling around the rows of barracks, three-storey red-brick Victorian blocks looming high over them. I was starting to feel the excitement rise as it dawned on me that one of those buildings would be my home for the next four months.

The bus pulled up by the Parachute Regiment depot, which was known as the Browning barracks. There was a massive aircraft, a Second World War Dakota, dropped outside it like a guard dog protecting the place. I'd been on three holidays in my whole life, each time to a caravan in Rhyl in northern Wales, so I'd never even seen an aeroplane before. Now here I was, about to join the Parachute Regiment.

We filed off and the nice guy who'd met us at the train station was gone; he was physically still there but he had magically morphed into a snarling dog. It was as though he'd switched the tape in his head and started barking at us,

'Right, you people fucking stand over there, you do that, you do that.' I was spinning, doing what I was being told, which was to form an orderly line with all the other lads, all new recruits, same as me – except they weren't the same as me. I was right near the end of the line, looking along it at them all, thinking one thing, *I am the skinniest person here.*

I was suddenly aware that this wasn't the cadets; I was now in a man's world. I'd grown used to being a face around Walsall over the past couple of years, someone people knew, maybe respected, but while I might have been a big fish in a small pond there, now I was in the wide-open sea. Some of those guys had big hairy chests, tattoos, fucking moustaches, for God's sake. Again, I started doubting myself, thinking, *I am out of my league here, I'm not ready for this.*

The big dog led us into the barracks and each of us was shown to where our bunk was. For the next six months, I'd be living in a large open-plan dorm with twenty other blokes. As well as a bed, each of us was given a locker and a kitbag. That was it. That was all we had in the world and this was all we could call home.

I was bunking next to a couple of Scottish lads, the first time I'd ever heard the accent, and at first I thought they were being really aggressive with each other, until I realised that's just how they talk. There was a posh lad opposite my bunk who said he'd been to boarding school on the Isle of Man, as if to remind me how uneducated I was. Everyone was presenting their best side, sizing each other up, wondering where they ranked, who they could beat, how they'd get on in a fight, trying to judge, trying to work out who was the best.

The next morning, I was given my kit, including my very first rifle. I can still remember it today: number 35 – an SLR (self-loading rifle), a proper elephant gun, and I loved it. We had no choice. We were drilled from the word go: your rifle is an extension of yourself, it will save your life, take good care of it. We were taken down to the gym to meet the quartermaster sergeant instructor, who like all the instructors was to be addressed as 'Staff'. All the recruits lined up again, me stood at the front, all the other big hairy bastards lined up behind me as this 6ft 5, triangular-torsoed human bulldog with muscles in his spit came snarling towards us. I was too terrified to take my eyes off him.

'Right,' he spat, 'how many of you piss the bed?'

That wasn't what I was expecting. I wondered if I'd heard him right but then he said it again. Was it a joke? I couldn't be sure so I tried to look behind me without him noticing, to see what the other lads were thinking, but as I got my head halfway round, I could see the big hairy boys had all put their hands up. So I put my hand up. That night, I called my mum from the payphone and the first thing she said was, 'Do you want to come home?' I didn't; I just wanted to tell her about all the lads who pissed the bed. She never even hesitated. 'Oh, so does your dad,' she said. *Fine*, I thought, *then I'm staying put.*

Our platoon was assigned our own dedicated PT instructor, who was horrible too and looked even more menacing than the quartermaster sergeant. He barked all the time and he sprinted everywhere. He wasn't our friend and he wanted us to know that. I don't think he had any friends. Anywhere. The gym sessions he ran were hard, but it was also the first time we got to see what everyone else could really do. Our first session was relentless. Relentless pull-ups, relentless push-ups – just my kind of thing, so I was actually enjoying it, especially as, when I looked around, I started noticing that some of the bigger guys couldn't keep up with me, and nor could the smaller guys. Everyone there was fit, for sure – really, really fit – but after years of boxing training, cardiovascular stuff was still my domain.

The Falklands War was only recently over and quite a few of our staff had served there, which meant the training we were getting was based on reality and experience – up-to-date experience. It was intimidating and frightening to think about how these men in front of us had actually been out there, that they'd probably had people killed next to them or even killed people themselves. They were hard and firm and I felt at the time that they were horrible and evil but, honestly, now I can see that they were great men. They inspired us, pushed us to the limit and, based on what they'd actually experienced, prepared us for what we would one day face on the battlefield ourselves.

As 17-year-old recruits, we couldn't help but imagine what battle was going to be like. A lot of the lads felt bad that they'd missed the Falklands, and they wanted to be there if it ever happened again. Everyone there wanted to do their bit and that was the reason most of them had signed up. You might be doing pull-ups in the gym, but going through your mind the whole time was, *I might have to shoot someone or they might shoot me.*

Operationally, Northern Ireland was really the only thing going on for British forces, and a lot of us knew that the Parachute Regiment had suffered over there. There were also certain IRA activities spreading onto the mainland, which some of us were aware of, but for the most part we were locked away from the outside world. We didn't get time to watch TV and the only communication I had with the world outside Aldershot was the ten minutes I'd get to call home and talk to my mum. I'd call her religiously every night, just to let her know I was all right (and that I wasn't pissing the bed). I called her every night because, really, I was still a kid at heart.

I don't think I was capable then of thinking about what it was like to actually go to war or to shoot someone. What I did know was that I wanted to go and prove myself somehow. I wanted to be really challenged, and war was what I thought of as the ultimate challenge – to be face to face with somebody who is probably going to try to take your life. I wanted to see how I could cope with that, see how I could deal with that ultimate win-or-lose situation. Training was a phase that I needed to get out of the way so I could get to the real challenge – join the battalion and hopefully go to war.

There were so many elements of the next six months that were hard: living out in the field was difficult, digging holes in the rain, total sleep deprivation, almost starving because all you had was rations, in the middle of the Brecon Beacons, which is harsh enough in the summer. I never knew how I was really doing either, because the instructors didn't give anyone a pat on the back and say, 'Well done, you keep going.' The only way you could judge how you were doing was by looking around and thinking, *Well, I'm still here so I must be doing okay.*

Because of all the training, I was finally starting to put on a little bit of muscle, starting to grow up physically as well as mentally. I was starting to listen more and, because of that, to learn more. Instead of being quick to make the smart comment, I was keeping my mouth shut. Instead of getting into fights, I kept my hands in my pockets. I knew that this was a different level now, a different league. These were serious people, unbeatable people, men who'd done serious things, and I felt a deep respect for them. I knew that they weren't going to take any bullshit from a mouthy fucker like me.

For the first time ever, I felt like I was in the company of men who'd proved themselves and been through proper hardship. It was the first time I'd ever had so many positive role models around me. Growing up where I did, I was used to people who'd rather create a problem than sort one out. With few exceptions, I'd only known men who were like me, scumbags, who'd cause trouble with somebody just to have a fight for no reason. Now I looked at the men I was in the company of and I started to see that I didn't have to be that kid who got into fights any more. I had another option; I could be more like them but, to do that, I had to change.

Change isn't for everyone. Seventy recruits quickly became fifty and then fifty gradually became twenty. We did weapons training, learned how to fight in patrol, how to practise aggressive action, how to do surveillance and build observation posts. I particularly enjoyed the counter-surveillance stuff, learning how to hide and gather information, having to whisper, always moving covertly, always staying camouflaged. There was a routine for everything: shaving; staying healthy; even for how to go for a piss. I'd never really thought about any of that before.

We'd get punished for doing things wrong, of course. The 'Firebreak' is legendary for the Parachute Regiment in the Brecon Beacons. From the parade ground, the 'Firebreak' runs 4km across the fields, through the break in the trees and up to the top of a steep hill. The instructors like it because you can be seen nearly every step of the way from the depot to the top. Every time we fucked up, we'd be sent off, rifle held high overhead, running the whole way there and back. There were a few dead spots, where you thought you probably couldn't be seen, so you'd bring your rifle down, give yourself a break. Then you'd get back and realise they actually had watched you.

'Right, I told you to keep that weapon above your head – did you do that?'

'Yes, Staff.'

'No, you fucking didn't, I watched you. Off you go again.'

It was all character-building, designed to develop your fitness, but was also a way of training your mind. I knew I needed to start thinking and absorbing information faster and sharper. I had to learn to tune in to what they were trying to tell me because my life might depend on it one day.

Training finally built up to a selection process – a three-day intensive set of tests called P Company. P Company was the single hardest thing I'd ever done in my life. We had to do a log run over a 2-mile course, as well as a steeplechase and an assault course, which was timed. The next day was a 10-mile run, carrying 55lbs on your back, in under one hour forty-five minutes. Then came the stretcher race, with 180lbs on a metal stretcher to replicate carrying a body off a battlefield, this time 8 miles and again timed. Everything was hard and marked out of ten. We had to get a certain number to pass. Final day was the 'milling' – a toe-to-toe fight with another recruit; one minute to show your courage and controlled aggression. To be honest, I felt sorry for the poor bastard who got selected to mill with me. After a life spent fighting and boxing, I was ready for that one. He took three punches before I sparked him out, so they picked a bigger lad for me to mill against. He also lost. And so did the next.

There's nothing showy about the P Company process. You survive it. After three days, I went back to the barracks and passed out on my bunk. I remember thinking, as I slid into unconsciousness, *I don't care if I pass or fail, I've done it now. And I'm never doing it again. Fuck that.* If they'd told me I'd failed then that was it, I was going home. I'd heard that there were lads who repeat P Company two, three, even four times, and now I admire them, but then I knew I wasn't ever going back and doing that again. No chance. When you're in the army, you get shouted at all the time, and I always hated it. Never got used to it, except for one moment. The day after P Company, all the recruits were summoned and we filed into the big hall. We lined up to attention and each one of us waited for our name to be called.

'Adlington?'

'Staff,' he replied, standing to attention.

'Failed.'

'Staff,' he said once more, stepping back again. Gutted.

'Billingham.'

I took a step forward, just the same. 'Staff.'

'Passed.'

A lump rose in my throat and I had to force the word out – 'Staff.'

That was it. I stepped back into line, trying to remain professional, but inside I felt like I'd just won the lottery, because that was the moment it happened. I'd passed. I was going to be a paratrooper. It was the greatest feeling of my life. I'd passed P Company and I fucking loved it.

Not that anyone who passed got much time to dwell on their success. The next morning we were rounded up, taken to Brize Norton and put in the hands of the RAF, which was a totally different experience. Nobody in the RAF is screaming and shouting at you. They're all nicey-nicey. Probably because they have to be – they're about to throw you out of a balloon.

Our balloon jump was scheduled for the following day. To prepare for a real jump from a plane, para training requires you to do a static jump from a Zeppelin balloon, hovering 800ft over Weston-on-the-Green. We bunched up on the green in groups of four. Me and three lads, wearing our kit, parachutes and all, knees knocking, were squeezed into a tight cage and waited for it to slowly climb up and up into the blue sky, everything becoming eerily silent the higher we rose. Suddenly the cage stopped with a jolt and one of the RAF boys unlocked the door, guiding us out onto the platform. I could barely concentrate on what he was saying as he started to explain the next steps. The one thing I can remember is him pointing: 'That's the cemetery over there,' he said with a big laugh. The RAF love to take the piss; just to make sure that if we weren't all shitting it before, then we were now.

He checked my reserve chute and nudged me on the back towards the edge. I had time for one deep breath before he tapped me on the shoulder: 'Jump.' Unlike jumping from a plane, there's no slipstream when you jump from a balloon; you just drop, like a funfair ride, except without the fun. Running through my head was the one instruction I remembered: 'Count to four.' If your main chute hasn't deployed within four seconds then you have to engage your reserve and, with only 800ft to play with, you'd better not get it wrong. That four seconds felt like four hours. But then *Boof!*, my chute opened and everything went quiet again. I was parachuting, I was actually parachuting. I could see the Oxfordshire countryside all around me, hear the song of the birds in the trees, a rare moment of peace at the end of the most intense four months of my life.

It was at that moment that I think I really started to believe in myself. I was in the Parachute Regiment, just like I'd set out to do when I left my parents' house in Walsall four months before. I had an intense feeling that I had joined the best of the best. I could look at all the other units, all the other members, all the infantry, and I could feel proud. The paras have a reputation for thinking they're better than everyone else. Because we are. Of course, I know people probably all feel the same way about their own regiments and a lot of them actually hate the paras, but, right then, I didn't care about them because I finally belonged to something special, something untouchable.

I'd achieved something and proved myself like I'd set out to. Maybe I wasn't a scumbag after all: maybe I was actually somebody worthwhile. I'd left behind that boy from Walsall who wasn't particularly pleasant, that idiot who was known for having a fight every week. I'd put my family through hell for long enough and now I was going to do great things, challenge myself honourably, prove that I could take this further, all the way. There was nothing stopping me now.

I was so happy. I had given myself a real chance and it felt great.

My mum and dad, sisters, brothers, aunties, friends, including Mac Gaunt from cadets, all came to the passing-out parade. A coachload of Brummies headed to Aldershot like they were going on their holidays. We did the balloon jump for them. Only, this time, we were all full of beans, showing off the new confidence we'd gained since that first jump. For the main parade, we performed the drill and all the recruits lined up along the parade square. I could see my mum and dad sitting right behind where the CO was addressing us all. The Parachute Regiment flags and Union Jacks were blowing in the wind either side of him as he called the regiment to attention. Then he shouted out, 'Men for awards.'

I marched forward fifteen paces with two other lads. We were stood front and centre as the CO announced what we were winning awards for.

'Private Billingham, champion recruit of 497 platoon.'

Seventy of us had started, seven of us had finished, and here I was – the best of the bunch.

I stole a sideways look towards my dad; he was sitting there, mouth open in stunned silence. That was all I needed to see. *I told you I could do this, Dad. I told you.*

I was happy that he was there. In all the years that I'd boxed, he never came to see me fight, but I think that was because he saw it the same way that I probably did. Boxing was for fun. It did help me, but it wasn't who I was. The army was different. By joining the paras, I'd marked myself as a man, a man with a career, a man who was to be taken seriously. My dad being there that day was his way of showing me that I'd done something he was really proud of.

I'm sure he wanted to shout out, cheer, clap, but that wasn't my dad's way. He was never the man who said, 'Well done' or 'I'm proud of you'. Instead, after the parade, he came and found me.

'Is the bar open?' he asked me. 'Can we go and have a beer?'

'Okay, Dad, let's go.' I knew what he meant. I knew what he was trying to say.

I left Mark Billingham the boy in Walsall. I was now Mark Billingham the man. I was in a man's world and I was going to do men's things – good, important, honourable things. That's all I wanted to do and I knew it like I'd never known anything before. I walked out of those gates feeling 6ft tall and thought, *The world's mine now, I'm going for it, this is fucking awesome.*

CHAPTER 4

This Is My Reality Now

When I heard my first posting was going to be to Belize in Central America, I thought it meant the middle of the US, somewhere near Kansas. It was my mum who first told me that Belize was a different country altogether and located in the jungle. Third Battalion were being sent to Belize as a visible presence in the war against narcotics, as well as to police some skirmishes that were going on with the Guatemalans along the border and to shut down some illegal logging and poaching.

In the months before we were due to leave, I was assigned to do duties in the guardroom back in Aldershot. It was an easy detail, designed to help me get settled in and learn the ropes of the new battalion. My supervisor was the guard commander, a guy called Benny Bentle, a big lad in his early thirties with thick dark hair. He looked menacing but he was soft-spoken and down to earth and was totally different to any of the guys I'd met during training. Benny had served in the Falklands and was then made the rear-party guard commander, in charge of the camp and all the prisoners.

Benny was the first soldier I'd met who didn't shout or bark at me. Instead, he explained to me what the Parachute Regiment was and how it worked, and how I should act now that I was part of it. He really took me under his wing. For hours, Benny sat and talked to me about the history of the regiment, the characters to look out for and those to be wary of, and about what to expect in Belize.

One of the characters he mentioned was a guy who was in the prison on base at the time called Ghost. Benny told us that Ghost was an infamous legend. 'Fucking do not upset Ghost.' I don't think I ever actually saw Ghost, but I didn't have to. I was petrified enough of his reputation, and the evenings when I'd have duties round the guardroom, I'd be thinking the whole time, *Please don't let him out when I'm here.* Benny told us that he was really a nice guy but someone had upset him. I knew the truth was that he was in jail for shooting somebody with a crossbow.

Benny really looked out for me during those first few weeks, reassuring me about how much I was going to enjoy Belize and teaching me to prep my kit, giving me advice on what I'd need or not need. He seemed so much older and wiser, like a proper grown-up. 'I'm getting married,' he said to me one day. 'I'm going to come out to Belize but I'm only going to do three months, just to make some money, because when I get back, I'm out of the army.' I was sad he was leaving but at least happy that he would be out there in Belize with us.

We flew across the Atlantic in an old VC10. Some of the lads I'd passed P Company with, like Tony Long, the guy I'd met on the train that very first day on the way to Aldershot, were now sitting next to me on a plane to Belize. It felt like it took about four days to get there, like I could've walked there faster. One hundred and forty of us, lads from all different regiments and a few from the paras, landed deep in the jungle just like my mum had said.

The door to the VC10 opened and immediately I thought I was going to die. As I stepped out onto the stairs and down to the jungle runway, I was smacked by the heat. The warmest place I'd ever been before that was Rhyl in north Wales. The humidity drenched me like I'd fallen into a bath, made worse by the fact that we were still kitted out in the British Army standard-issue wool shirts, which were instantly saturated, and mine started hanging off me like it'd grown two sizes.

I was struggling to breathe as I tried to take it all in. I knew I was fit, but I thought then there was no way I was going to survive living out there. It was like walking into a sauna that had been left on full blast all day. The dirt was bright red, there were palm trees and noises of insects and animals in the tree line that I didn't recognise, plus the overbearing smell of sweaty men's bodies hung in the air. It was a total sensory overload.

We collected our kit and got loaded onto an old battered bus to be driven to our billets. Nobody was talking, even the other three lads I knew from 3 PARA were lost in their own heads. It was like joining the army all over again, each of us looking around, trying to get our bearings, trying to make sense of it all. We were all in the same company but assigned to different platoons, which meant that we could see each other in the evening but during the day we'd be doing different things. I wouldn't see my mates every day because my platoon was stationed on the other side of the airport camp, on the main base next to the Harrier jump jets.

Our digs were in what were known as the 'Laundry Lines', a collection of Nissen huts next to where the camp's washing was done. I stepped down off the bus, back into the sweltering heat, and lugged my kit past the line of small, dark Central American women, crouched over tubs of suds, peering back at me as they scrubbed the dirty uniforms. They were the first foreigners I'd seen, except of course this was their country, which meant I was the foreigner.

I was given live rounds for the first time and, like everyone in the company, was assigned kit to carry. Of course, as the new boy, I got given the biggest gun because it's the heaviest – the LMG (light machine gun), just about going out of service. Then, later, I got given the 84 – the Carl Gustaf, an 84mm anti-tank weapon. Weighs a ton.

In the first few days, we were assigned to conduct border patrols in the south of the country. It was my first time jumping with the battalion, but now, instead of ten of us in the plane, there were ninety of us. Of course, I was put right at the back, bouncing around, trying to fix my anti-tank weapon to my leg and hold my rifle, with kit going over everywhere. One of the lads next to me was sick. I knew the guys in the company were sizing me up. I could feel them watching me, seeing how I'd react, but I wasn't going to let myself down. This was a chance to keep my head down, get on with it and prove myself to them.

On our first insertion into the jungle by helicopter, we landed in a clearing just back from the border, tasked with looking for illegal logging gangs using drug hides and flushing out anybody who might have been hiding there. Despite the heavy guns I had to carry, I was beyond excited, listening to all the guys around me telling stories about what they'd done and thinking that I couldn't wait to get stuck in.

Our local guide was an expert in that area of the jungle. George was half Mexican, half Belizean and he led the squad like a scout. I, of course, was at the back again, doing as I was told, keeping quiet because I was new and I hadn't learned yet how to navigate in the jungle. Plus, I had the fucking house on my back as well as the anti-tank gun and my rifle, so I needed to save my energy.

We quickly shook out into a patrol, following George's lead, no one really navigating because George knew the jungle like the back of his hand and any time someone stopped to get the map out, George would tell him to put it away – 'no problem, no problem' was like a catchphrase of his. I loved it; I could hear the sounds of animals and birds I'd never heard before, half expecting a tiger or an elephant to jump out on us any second, even though all I ever saw were snakes and spiders.

I loved the noise and I'd settled into the heat. After a few weeks, I didn't mind the conditions any more. I didn't mind being wet and sweaty and dirty, and I loved the smells and the relentless buzz of the forest. It's claustrophobic and it's noisy but as soon as you can get used to that, it almost becomes tranquil. Some people couldn't bear it; people freak themselves out because navigating in the jungle is hard. But I enjoyed it. I enjoyed the challenge of not being able to see things from a distance. There's just something about it that, to me, felt peaceful. Maybe it was because I was a kid from the urban sprawl of the West Midlands, but I just really loved being in nature, so close to everything.

Night falls fast in the jungle and, as it started to get dark, the troop sergeant asked George where we should make camp. 'Not my fucking problem,' was his reply. I had to make sure I'd heard him right. He shrugged his shoulders. He clearly had no idea where we were. I was at the back chuckling to myself because we were now in a swamp, which meant we were going to get eaten alive all night. It seemed to release a pressure valve somewhere inside of me. People could make mistakes. It wasn't all serious, all the time.

I knew enough not to complain about that kind of stuff, too. I was just happy to be there. It was hard, of course, and intimidating, because all these big men were twice the size of me. But I kept my head down and got on with whatever I'd been tasked to do. That's how I settled in. The other lads accepted me because I was having a chuckle with them, carrying all the kit and not whingeing about it. I didn't mind doing it because it was all part of showing these guys what I could do. This was how I would become one of them. There was a reason I was at the bottom of the pecking order and I accepted it. I was at the back because that's where I deserved to be, for now. I knew that one day, when I'd earned it, I'd be at the front.

Looking back, Belize was the perfect introduction to the army. A first patrol in Helmand or downtown Baghdad would have been so much worse. Belize was classed as operational and there were genuine incidents going on. One patrol shot a couple of guys who were smuggling drugs over the border and had opened up on them. There were skirmishes we heard about involving other companies, too. It felt real, even if I didn't have to engage an enemy or come under fire myself. Every day was exciting and challenging while giving me time to settle in and enjoy being a part of it. When our section wasn't on patrol, we'd be on camp security – manning the gate or the guardroom, or doing lookout up on the sentry posts. The rare times I got to myself, I spent training, keeping fit. I'd run circuits round the camp, swim in the pool or hit the gym to keep myself in shape. Fitness was always important to me.

A few weeks in, I was down at the swimming pool, training with some of the lads from A Company, when the Tannoy went off: 'Stand by, emergency.' Immediately the helicopter came in, low and loud over the jungle, and landed outside the main medical centre. I ran over to see for myself what was going on, watching as they pulled a covered stretcher off the helicopter and laid it on the ground. One of the medics pulled back the cover and I just saw it. It was Benny. He'd only been out a couple of weeks, doing training exercises with live fire, when somebody had shot him by mistake and killed him.

My whole world just crumbled. Benny was the guy who had guided me, talked to me, really the first person in the regiment I could say I knew, and now here he was being brought off a helicopter, dead. Going through my mind was everything he'd told me about getting married, how he was leaving the army, how he'd survived the Falklands, all that he'd done and now he was gone. It was horrendous, but also a serious wake-up call. In my mind, all I could think was, *This is real. This is no joke. This is my reality now.*

CHAPTER 5

I've Just Seen You Do That

I feel bad because the way my story sounds up to now, it's as though my whole life was spent fighting. I suppose a lot of it was, but by the time I was in the Parachute Regiment, I felt that things were starting to turn around and I began to feel different. I was now a soldier and soon-to-be husband, not that wild, rogue person any more. I was a man; I'd proved myself as a worthwhile person. I was ready for the next steps, to have a wife and hopefully build a family of my own. Soon after I joined the Parachute Regiment, Julie came down to be with me in Aldershot and we got married. It's a hard life being a soldier's wife. I didn't get paid a lot so we didn't have much money but, despite that, it was an exciting time.

The army gave us our first house together, albeit in the tenement blocks, near the camp. They were so rough that I was delighted when I found out we were going to be moving in at night-time, because I'd been worried Julie would burst into tears when she saw how basic it was. But we tried to make the most of it, we all did. We weren't alone, because a lot of the other lads who'd joined with me were getting married and settling down, too. There was a nice community, lots going on; we were making a life for ourselves. Not any kind of luxury, but we had an opportunity to settle into a new life and be a real couple. Unfortunately, we had to put that on pause, because a few weeks after we moved into our new life, I got the order that I was being deployed to Cyprus for six months.

Everyone knew what a deployment to Cyprus meant. It meant beach, sunshine and the nightlife of Ayia Napa. It was known as the dream deployment, like a reward for completing a more arduous assignment elsewhere. We'd worked hard in Belize, completed more training in Scotland, and so the lads and I all saw this as our reward. I couldn't wait to get out there and let my hair down. Cyprus is divided right along the middle by a border with a buffer zone on either side: Turkish to the north, Greek to the south. The conflict there has been going on since 1974, when Turkey invaded and occupied the northern third of the island. Then, in 1983, the Turkish community took it a step further when they declared independence and formed their own Turkish Republic of Northern Cyprus. Even though nobody except for Turkey recognises it, they've been there ever since.

The UN stepped in to try to help everyone find a peaceful solution and came up with the idea of the buffer zone to keep the two sides apart. Although they tried to have talks to bring the two sides together, things have never been worked out between them and every now and again hostilities resume. Our job, while we were stationed on the island, was to simply keep the peace and make sure things didn't kick off again.

For the first three months of my deployment in Cyprus, half of the battalion would be in Dhekelia, training and getting their skill sets up to speed, while the other half were stationed nearer to the capital, Nicosia, manning border posts along the buffer zone – a 180km-long track dividing the island in two. I drew the short straw when I was part of the team put to work. The first sign that my 'holiday' posting might not be all I'd hoped for came as soon as we landed in the sweltering heat, to discover that the PLO (Palestine Liberation Organization) had just upped their operations against UN targets. There was already a rumble in the air that something bad might happen and our intelligence said that the PLO were shifting much of their focus from attacking civilians to more military targets. The first night after we arrived there was a mortar attack on Akrotiri. The mortar was fired right into the base and was followed by small-arms fire on the perimeter. The following week there were another couple of shootings.

We set about strengthening defences, building road blocks along the main roads, stopping vehicles and conducting searches – more the kind of thing I'd have expected to be doing in Northern Ireland than outside tourist resorts in the Mediterranean. News came through of a soldier from another battalion who'd been shot dead while driving a vehicle. Then another story came in of a shooting near Larnaca down the road, another guy killed. Far from a holiday deployment, I'd landed in the middle of a full-on security-risk situation and, days later, the warning level was set to red alert. Instead of dancing on the floors of Ayia Napa, we were running around the streets trying to find lunatics who were hell-bent on killing us.

One good thing to come from that posting in Cyprus was that I learned how to conduct a patrol. We patrolled each day in groups of eight, spilt into two groups of four. Both groups had comms with each other and the squad commander had comms back to base. Our tasks varied from vehicle searches to surveillance and street patrols within an area of 4 square kilometres for three or four hours every day. As well as the comms, water and medical kit, our troop had a heavy machine gun or two, maybe a 66mm anti-tank weapon, our own rifles, grenades and live rounds. I arrived in Cyprus in May, so it was a shock to find myself carrying all that weight in 30-degree heat. A patrol is only as effective as the people on it. It's about professionalism and concentration, and a good commander knows how to keep his people focused on the job in hand. But in Cyprus our CO had his work cut out because every day was exhausting; every day we were on edge. The adrenaline I felt on patrol was an unusual paradox because, while we were there to ensure that nothing happened, I still found myself almost wishing something would, just to release the relentless tension. I was willing the PLO to come and attack us so that I'd have someone to target, some way to bring an end to the constant expectation of an attack.

I don't think we achieved anything in Cyprus other than more soldiering. I improved my skills, learning how to cope with a different climate, which was good for me personally but hardly the point. I'm glad I went and experienced it because it was a wake-up call for me that not everything is black and white, not every problem can be solved. The situation in Cyprus now isn't that different from what it was then. The whole exercise was and continues to be a massive waste of time and money but that's just how it is. It wasn't the only time I'd find myself frustrated by how pointless a UN mission would feel.

The six months in Cyprus dragged on and, by the time it came to an end, I couldn't wait to get home. Not least because, just before I'd left, Julie had fallen pregnant. Shortly after I returned, I was going to become a father. In between operational tours, the Parachute Regiment sent me around the UK on airborne training exercises. Almost as soon as we returned from Cyprus we began going on exercises, always as part of a team on a scenario – could be surveillance, could be an attack – the idea being to replicate as closely as possible what might happen for real in a foreign land. Operations were usually airborne, which meant they began with parachuting into a 'hostile terrain' and then role-playing the operation on the ground for a week or two.

I was in Scotland on one of those training exercises, in the freezing ice and snow for two weeks, when I got the call to say that Julie had gone into labour. Immediately my CO told me the news, I was scurried away from the battalion, thrown into the back of a truck, and we bombed it down the M1 to Aldershot. As we drove towards England, I could feel that I had a bit of frostbite in my feet from being out in the snow so long. The pain was excruciating and got worse as my toes began to thaw out, so by the time I got to the labour ward, I could hardly walk. I arrived at the hospital just in time to see Julie give birth to my eldest daughter, Zoe. She was already mid-delivery when I stumbled in and saw the midwife giving her gas and air. I took one look at it, grabbed it off her and started taking it myself. I needed it more than she did.

That Christmas we went back to Walsall on leave with our new baby, and I went out on the town to see some of my mates and wet the baby's head. A whole group of lads, who I'd known since school days, were out and the pubs in town were packed with people getting drunk, getting in the mood for the holidays. There were two pubs in particular, right opposite each other – one called the Black Horse and the other the New Inn. They were both pretty horrible pubs, but that night they were where everyone had headed to. Before long, they were so packed that people were spilling out the door to drink in the street.

I'd had a few beers and I was relaxed, joking with my mates, telling them some of the stories about what had been going on since I'd last been home. There should have been a good feeling in the air but I started to get a sense that something wasn't right. There was a bit of an uncomfortable atmosphere in the Black Horse, so we decided to leave and go over the road to the pub opposite.

While I was waiting at the traffic lights to cross the road, a bit of a commotion started kicking off across from me. A group of lads I half recognised, a few of the old faces, people I knew as having a bit of a reputation and who probably knew the same about me, were looking at us and shouting some nasty stuff our way. Before I knew it, they were tearing across the street towards us, backing us away from the road and towards the shops next to the pub. An almighty fight broke out; twenty lads kicking and throwing punches, stamping and whacking each other. Suddenly two lads grabbed me and, in the scuffle, we all went flying through the plate-glass window of the butcher's shop. I lay there showered in glass for a moment; it was a miracle I hadn't been cut to pieces and I climbed back out onto the pavement with just a few scratches. I stumbled back into the fight, which by now was happening out in the middle of the street. Suddenly, I heard an almighty, deafening screech of tyres and then a thud. I watched, time frozen, as my mate, Bani, went flying 20ft through the air right past my ear and landed on his head.

Before any of us could react, the sirens were blaring and the street was filled with blue lights. We all scattered, running from the danger. I was suddenly sober, suddenly aware of the potential consequences of what I'd just been involved in. What the fuck was I doing? If I was arrested for fighting in the street, it would jeopardise everything I'd just worked so hard to achieve in the past year. I had responsibilities now. For the sake of a pointless scrap, I could have just thrown away the future I'd been slogging my guts out to build for myself, a proper career, a chance to make something of my life. I started to run. I ran like I'd run when I was that 9-year-old kid, but this time I was running home. I ran down the hill and over the bridge, halfway to the end of my street. All I could focus on was getting home, shutting that front door behind me and putting this whole stupid mess out of my mind.

I was over the bridge and about to turn off the main road when I saw a police car bombing up the road towards me. It rode up the kerb, blocking my way, and two coppers bounced out. I had nowhere to go so I held up my hands to show that I meant no trouble. *Whack!* I took a truncheon to the back of the legs and fell forward onto my knees. The second copper came in from the side and, again, *Whack!* This time to the head, throwing me forward onto my hands. I knew the game was up. I was done for. This was the end of everything and I was going to take a beating, too. What a bloody waste. Just then, I heard a banging sound. The copper stopped hitting me, looking up to see where the sound had come from. I turned my head to see the face of a little old lady in the window of the house right next to us. She was hammering on the window with everything she had, which made the coppers as surprised as I was. We were all frozen for a second, watching as she opened her window.

'I've just seen you do that,' she roared, shaking her fist.

The coppers looked at each other like naughty schoolboys, caught getting up to no good.

'Where are you going?' one of them asked me.

'I'm going home,' I said.

'Right, make sure you do.' He turned on his heels, the two coppers jumped back into the car and they tore off up the hill.

I got back to my feet and nodded to the old lady. She had no idea what she'd just done for me, but, in truth, she'd saved me. I ran all the way home, ran back into the house and sat down on the sofa, head spinning, thinking, *What have I done?* I didn't know if Bani was dead (he wasn't, he recovered), didn't know why the fight had even started, but, most of all, I didn't know why I'd allowed myself to get involved in all that shit all over again. I couldn't go back to being that wild kid from Walsall who spent his life fighting. I was better than that. I went to sleep that night resolved that that was going to be the last stupid, pointless fight I was ever going to have. From that moment on, I was going to stay out of trouble.

Over the year that followed, we returned to Aldershot and I kept that promise to myself. I kept my head down and concentrated on developing as a soldier and becoming a reliable father and husband. By the end of the year, I discovered two things: one, Julie was pregnant again, and two, that I was being deployed on my second operational tour. For the next two years, I would be stationed in Northern Ireland.

—

The British Army was initially deployed to Northern Ireland in 1969. I was only four years old when the Unionist government of Northern Ireland requested support from the British government in London for the Royal Ulster Constabulary (RUC) after three days of rioting across the province had gotten out of control. The army responded, putting boots on the ground to keep the peace, and continued to have a considerable presence in Northern Ireland during the conflict that followed, known as the 'Troubles'. At its peak, there were over 20,000 British troops deployed, and before the Belfast Agreement finally brought peace in 1998, over a thousand British military personnel had been killed.

As I have said earlier, my mother was a Catholic and my father was a Protestant, not that it really mattered in our house. Neither of them ever made a big deal out of it, and none of us kids were really raised in either faith, so when I heard I was going to Northern Ireland I wasn't particularly concerned about the politics of the situation. My parents were worried that I could get blown up, injured or killed, but it was just another posting as far as I was concerned.

Our role in Northern Ireland was simple: we were there to support the work of the RUC, giving them protection while they were carrying out normal policing duties in areas where there was a perceived terrorist threat, or patrolling around military and police bases to deter terrorist attacks. In the event that the RUC had reason to believe there was specific terrorist activity, then we were also there to support and direct any necessary counter-terrorist operations. As I would be stationed at Holywood barracks for two years, it qualified as a family deployment, which meant that Julie and Zoe came with me. We were assigned our own little house on the camp; Zoe went into nursery and, shortly after, Kayleigh was born in the hospital in Dundonald.

Bombs were going off in and around the place all the time. One night the windows of our house actually bowed in with an explosion that went off in the city a couple of miles down the road. We'd put plastic on them to stop them shattering but, still, everything on the window ledge went flying across the room onto the floor. It was hard because my instinct was to stay and make sure that my family was all right, to protect them, but I also had duties as a soldier – that was why I was there, to protect the people – so an explosion like that meant I had to go back on duty, to take care of the fallout that it created.

Our tasks included running vehicle checkpoints, house raids, stopping riots and supporting the RUC on arrests. The very first time I was sent out into the streets of the Ardoyne, I could feel the tension and the animosity being directed towards us and we'd regularly get bricks and bottles thrown at us during patrols. I thought it was odd that we were sent in wearing our berets. The Parachute Regiment's maroon berets were well known, especially after its involvement in Bloody Sunday, in 1972. As far as I could see, the maroon beret was now an inflammatory symbol, unhelpful to the situation and a potential source of further provocation.

Looking back, I think even then I realised that we were on thin ice in Northern Ireland. I'm not sure what right we had to be there, getting involved in people's lives. Of course, we were looking for terrorists, trying to keep the peace, but

I could understand how, for those who had nothing to do with the 'Troubles' or the politics, we were seen as a nuisance. So I tried to remain as human as I could all the time. I was clear in my head that I was there to try to make people more safe, whether they resented me for it or not. One night, while we were manning one of the RUC checkpoints, searching vehicles, looking for weapons, we stopped a bunch of young lads. I asked the driver if he had been drinking.

'No,' he said, but I could smell booze coming off him.

'Well, in that case you won't mind blowing into this then,' I said. Now, I'm a soldier, so I no more had a breathalyser on me than a unicorn. What I did have was a night-sight off my rifle, so I had him blow on that.

'If that goes green, then you're in trouble,' I said, trying to keep a straight face. Of course, the second he blew into it, I switched the light on and it went bright green.

'Fuck, okay, right, I've had one drink,' he said, before me and the other lads on the checkpoint burst out laughing at him.

'Go on,' I said, 'get on your way.' And we let him off.

I like to think that moments like that had more of a positive effect on relations between the two sides because, at the end of the day, we're all the same, and having a bit of a laugh together is a more powerful way to bond us than anything else I've ever come across. But there was a serious side to why we were there, too. I was assigned to the Close Observation Platoon, carrying out surveillance on known terrorist players and covert searches, looking for weapon hides. If we heard there were weapons being stored in a house or a barn, then we'd go and take a look in the early hours of the morning, taking care not to spring any booby-traps, checking out if the intel was true or not. The idea was not to cause trouble if we didn't have to.

Sometimes we'd follow suspected terrorists, from both sides of the conflict, watching them from covert locations, marking out their routines to see how active they might be or whether there was something different in their behaviour that required attention. We needed to be sure that, if they had to be picked up, we knew the best time and location to do it. We'd hide in the mountains, in bushes or in a derelict building opposite where they lived, taking long-range photographs of them and the people that they hung around with. Gradually, we built up a profile of who they were and what they were up to.

My time in Northern Ireland was interesting for me personally because I found that I enjoyed working covertly. I was good at it. I liked the challenge of operating under the radar. Why not gather intel on dangerous characters with the minimum of fuss when we were in such a potentially explosive scenario? I didn't see the guys I was profiling as soldiers; I saw them as terrorists, cowards who would rather hide and blow up civilians with a bomb than stand and fight their enemy. Anything I could do to help apprehend them and get them off the streets was a good thing.

I grew massively during that time too, because it was all very real, a world away from simulated training exercises.

The news came over one afternoon that our battalion had lost three guys, blown to smithereens on patrol down at Mayobridge. A lad I'd boxed with in the gym and liked a lot, Spike, was killed by a joyrider on the Falls Road while manning a checkpoint. Since Benny, it was the first time that people I knew well had been killed. It made me aware that I could be talking to someone one day and then holding bits of them in my arms the next. I knew I had to step everything up another level. I had to have eyes in the back of my head, always be aware of the danger that could be around me.

I was manning a sangar in a static position overlooking an observation post one night on Templar Avenue, not far from the Falls Road. We'd had some intel that the IRA were planning to attack one of our observation towers. We already knew that their technique then was to put bombs in drainpipes and then push them up and underneath the platform where the soldiers would be standing, so I'd been given the job of running observations on the drains.

Nothing happened for days until one night I saw a black glove come over the top of the observation post balcony and then a head pop up next to it. The figure was wearing a black balaclava, looking around for somewhere to put the bomb he was carrying in his other hand. I reacted right away, trying to get an aim on him, trying to see if he had a weapon – he didn't, which meant that I couldn't fire. All I could do was sound the alarm but, by the time the quick-reaction forces arrived, he'd disappeared. My heart was pounding. It was the first time that I'd come face to face with a terrorist.

After a year, I was promoted to corporal, the first time anyone had ever given me a position of responsibility in my life, and I was offered a posting back at the depot in Aldershot to work as an instructor. It's a great honour to be given a chance to select the new generation of lads who will make up the new Parachute Regiment, but still, at first, I didn't want to go, thinking that I'd prefer to see out my two-year tour with the other lads I'd come out with. But I also knew that it would help my career if I took the promotion and the job in Aldershot.

I was growing up. I had a wife and two kids now and I had decided that the army was going to be my long-term career. I knew it was where I belonged and so I needed to see it that way. I'd also begun to think about joining the SAS, but I knew there was a two-year waiting period before I could even apply for that. Two years at the depot would allow me to get my skills up and make sure that I was as fit as I could be before I went up for selection. Everything pointed to the right decision being a move back to England.

—

The posting as an instructor back at the depot in Aldershot was a two-year job and I knew that, when it came to an end, I would be cycled back round onto operational duties, which probably meant Northern Ireland again and maybe Cyprus, too. Only this time, I would be going as a sergeant.

But something about the idea of promotion was bothering me. Sergeant is a vital role in the military, but it is a more strategic role than corporal. I was still only twenty-six and wasn't sure that was a role I wanted yet, because I was very happy being a corporal. As far as I was concerned, corporal was the best job in the infantry. It's where the action is; you're a commander but you're still in among it. After two years at the depot, I felt that I would be ready to fight again, and I was craving that sort of action, but I suspected that I wouldn't get it if I made sergeant.

Another thought had also started to take hold; a seed of an idea that was beginning to germinate. I knew I wanted to have a go at applying to join the SAS, and I knew that, by the end of two years training new recruits every day, I'd be really fit, so probably as ready for it as I'd ever be. A lot of my mates had already gone there after Cyprus and Northern Ireland and, although they didn't really talk about what they were doing there, I knew they were having a good time. I'd often read in the paper where the SAS had supposedly been and find myself wondering if that's where my mates were. I started to feel that's where I needed to be too, so, finally, I decided, *I'm going to do this.*

Nothing in the military is ever straightforward. Our group at the depot was pretty tight and the lads talked openly about what they wanted to do next, and where. Before long, it was no secret that I had ideas of applying for SAS selection, and so I didn't expect any push-back when I decided to formally request permission. However, my CO had other ideas. I applied for a place on summer selection in August 1991, but my request was denied. My CO wanted me to see out not only the two years of my posting, but another six months, then, in return, he'd allow me to go in the winter of '92. I was fucked off. I felt like I was being held back and I didn't want to wait, didn't want to stay in Aldershot when my mind was set for the next challenge. But I had no choice, so I decided that if I was staying, then I was going to make the time count.

I was promoted to the special training wing, which meant that I had responsibility at the depot for higher-level medical and navigation training. It also meant that I had more time, so I focused it on getting ready for selection. I was already fit, because as a Parachute Regiment instructor, I took my platoon of new recruits for a 10-mile run every two days, carrying weight, running with them every step. But now I started volunteering to go out on other instructors' runs, too. I'd go out with my own platoon in the morning and then I'd go out with another platoon in the afternoon until I was doing two 10-milers most days. I think the recruits must have thought I was a madman.

Another six months in the depot had other benefits, too. I got to spend more time with the kids. Working in Aldershot meant that I had something more like a normal life for a while. In the evenings, I could go home to my family, spend time with the kids, put them to bed and do the ordinary stuff like a regular dad does. Julie and I could do things a couple would do, like go out to the pub or take the kids out for the day. Although the job still required long hours, it felt more balanced and gave me the chance to learn what being a father meant.

My own father had been a real ring of steel around us when I was growing up and I wanted to be able to be that for my kids, too. My dad had grafted all hours and sometimes that meant he wasn't around as much as I would have liked him to be, or he was knackered and simply didn't have the energy for us. I wasn't in that boat because I never had to work half as hard as my dad. Of course, I'd be away on ops or training when I couldn't be around for my kids, but when I was back I wanted to use that time to be with them as much as I could.

I'd always been a naughty kid, and so it's probably no surprise that I fell into being a naughty dad, too. Cleaning out some of my old stuff one morning, I found some simulated grenades in the bottom of a kitbag. We used to call them thunder flashes in the army. I thought, *Fuck, I shouldn't have these. I'd better get rid of them.* So I sneaked them into a bag and out of the house one day when I took the kids down to the park. It was about five o'clock in the middle of winter, so it was already getting dark. I had the idea to put some weight on the bottom of the grenades, Sellotape them together and then throw them into the middle of the duck pond, where they'd sink to the bottom, never to be found. There was no one else in the park to see us, so I reasoned that would just make the problem go away.

It was one of those typical crisp British winter evenings and the kids were happy because they always loved seeing the ducks and running around the pond. So while they played, I strapped up the grenades and, when I was sure there was no one around to see me, quietly flung them into the water.

BOOM! I'd assumed the pond was deep. It wasn't. It was 2ft deep at the most and the grenades hadn't had enough water to softly sink to the bottom as I'd thought they would. Instead, they'd exploded with full force, right in the centre of the pond.

The kids stood frozen to the spot as a huge fountain of water rose 20ft high up into the sky and came down drenching us all from head to foot. All of a sudden, lights started coming on in the houses round the edge of the park as people reacted to the huge explosion they'd heard outside. I grabbed the two kids, one under each arm, now screaming because a couple of the ducks had keeled over, and we ran out of the park. We ran all the way home, our wet hair sticking up, looking like lunatics. Julie took one look at the three of us and said, 'What the fuck have you done?' It was a happy time.

It's sometimes hard to reconcile being a dad with being a soldier. The job is dangerous and even more so in the special forces, so my decision to apply for the Special Air Service, or SAS, wasn't taken lightly. On the plus side, I knew that at least in the SAS I'd get genuine downtime between operations. There wasn't much time off in the Parachute Regiment and deployments tended to be medium- to long-term. In the SAS, I would get paid more, so we would be able to go on proper breaks together, and we'd move as a family to Hereford, which is a beautiful place to live. Julie always said that whatever I wanted to do, she'd follow me and go with it. She was true to her word. We decided that I should give SAS selection my all and, if I got it, then we'd all get it.

Right before Christmas, the paperwork came through confirming that I'd been approved to attend the SAS selection in the winter of 1992. I was fucking chuffed; this was the chance that I'd been working my arse off for. I'd prepared myself physically and mentally for the challenge that was coming and I felt as ready as I'd ever be to take it on. First thing after New Year, I took all my equipment to be boxed up and stored in the regiment quartermaster's stores, in case I came back with my tail between my legs the following week. I said goodbye to Julie and the kids. If I made it through the first stage of selection, then we would have no contact for the next month.

There were three other guys from the Parachute Regiment I knew who were also having a crack at selection and one of them had a car, so the following morning before dawn we piled in and drove the 150 miles from Aldershot to Sennybridge camp in Brecon, each of us feeling the nerves, feeling the anticipation, daunted by the size of the task that faced us. There's a rite-of-passage element to paras having a go at selection. It's not for everyone, and some paras are paras through and through, but, still, probably more have a go at SAS selection than not, so all the guys in that car felt like this was something that they had to do. It was inevitable that they would be making the trip at some point in their careers. Still, we all knew too that going up for selection is very public; we'd told everyone back at Aldershot that we were going off to join the SAS. People judge, and everyone who fails has to come back and face that judgement.

A couple of years before, a friend of mine, Trev, had gone off to join the SAS. Before he left, he gave us all the list of reasons we'd never see him again – he was off to bigger and better things, he was destined for greatness, made for the SAS and all the usual bravado. Only he was back three days later with a different list, a list of excuses for why he didn't make it – he didn't like the directing staff (DS), he hated Hereford, he was a paratrooper through and through, the SAS were all crap-hats. I called it 'Trev's Ten Tantalising Excuses'. He had to take the ribbing from me and everyone else, because that's how it was, even if it wasn't pleasant. Driving down to Herefordshire that morning, all I could think about was how nobody would ever want to be in those shoes, especially not me.

CHAPTER 6

Selection

We parked up and made our way into Sennybridge to present ourselves for eight o'clock as instructed. There were men from all over the military arriving at the same time as us, dressed in uniform, with all the different-coloured berets denoting their regiments. It was as if the whole of the fucking British Army was represented on that parade ground. I could see a few more familiar maroon berets belonging to lads I knew from the Parachute Regiment, and there were men from the infantry, marines, navy and even a couple of RAF pilots. In all, there were 183 hopefuls, lined up in free rank, kit out in front of us along the parade ground, ready to try to join the world-famous SAS and all looking as confident as I was, when underneath the bravado we were all shitting ourselves as to what was coming. The not knowing played on all our minds.

But before any of us got anywhere near thinking we belonged, we had the small matter of months of selection to get through, and we all knew the odds were stacked against us: out of the couple of hundred of us stood there, probably no more than ten at best would make it through. I was immediately cast back to that first day in Aldershot when I'd joined the army as a wild teenager. Here I was again, standing to attention, preparing myself for a seriously tough selection test, looking along the line at the men I would be competing against. Again, my first thought was how much smaller and leaner I was than most of them. Again, I was asking myself what I'd done, questioning what I was doing there, the voice in my head berating me, saying over and over, *You've no chance, no fucking chance.*

It occurred to me that I'd been on both sides of this experience now. I'd done it once before as the person being selected, then I'd done it again as the person doing the selecting. Now, here I was, back in the vulnerable position. I almost chuckled as I realised that I was doing this voluntarily, trying to prove something to myself again. Only now it was a completely different league. The Champions League.

I'd been preparing for this moment for years. I knew I was more competent as a soldier; I was fitter than I'd ever been and I now knew better what the army was about. I had made the right choices with my career, taking on operational tours, gaining promotions, honing my skills and my fitness as an instructor back at the depot, but still none of that stopped me judging myself against other people's appearances all over again. It didn't stop me thinking about how big the lad next to me was or how fit the one over there looked.

One by one, each man's name was called and we were assigned to billets in groups of thirty. Our dormitory was fifteen double bunks and I was up top with a lad called Bellamy from the infantry below me. After orientation, we signed the necessary paperwork and passed the medical, and then our instructor asked us to prepare for a physical in the afternoon. I say asked because that's exactly how he communicated with us. It was one of the first major things that struck me, especially having been an instructor myself for the past two years, how politely and reasonably he asked us to get ready. It was almost eerie not to be shouted at, not to hear orders barked at me. In the paras, I regularly chewed out recruits as a matter of professional pride if I thought they weren't meeting the standards we demanded that I could see they were capable of if they put in more effort. Some lads couldn't handle it. But the SAS is different and, even on day one, there were early clues to just how different.

The physical that afternoon was a run. The rules allowed everyone to dress how they liked, so some lads turned up in the special boots they preferred or their favourite poncho or lucky shorts or whatever. I just went with army standard issue. I knew I was fit because I'd been preparing for months, years even, but no matter how fit I was, the guy taking us for the run would always be twice as fit. And he'd have the psychological advantage of being at the front, the only one to know how far he was going to go and how fast. We had no idea. Plus, my mind was playing tricks on me, because I was competing with the guys in my group as well as the ones who went in the first squad and the group coming behind.

From the off, it was hard. I was sprinting and asking myself what was going on. This was just a run, in sports gear, but after the first few minutes I was already thinking, *Fucking hell, how fit are these fuckers?* Again, I started doubting myself, wondering how I was going to survive this. I finished in the first group of ten out of forty, but I wasn't happy with myself. I'd had to work way harder than I should have done to achieve that. It was still day one, it was a warm-up day, a chance to settle in, and I already felt horrendous. I put it down to nerves and tried to get my head down. There was already a lot of apprehension flying around the dormitory that night because everyone knew that day two was the 'Fan Dance' – a gruelling 24km march, starting at 0500 hours, over the highest mountain in the Brecon Beacons. It happens on day two to sort the wheat from the chaff nice and early, because anyone who doesn't make the time is on the bus home. No excuses.

The guys who ran selection were the directing staff, addressed always as 'Staff'. The night before the Fan, the DS came into the dorm to talk to us, again being spookily nice as he explained that tomorrow we would each be required to carry a pack containing 60lbs – not 59, not 61, but 60lbs. Other than that, we were free to dress as we wished, but the march would begin at 0500 sharp. He told us all to get a good night's sleep and wished us luck and, the way he said it, it sounded so easy. We were being treated like grown-ups, which felt surreal and a little worrying. Were they trying to catch us off guard? Every man in our dorm knew he was capable of doing that run in normal circumstances – some had been up and down the Fan a few times already to rehearse it, so they actually knew they could do it – but still every one of us also knew that everything is different on the actual day. It didn't factor in the pressure that competition brings. We all knew we had to be at our very best.

In the middle of the night, I woke up. Someone was moving around below me. It was Bellamy. I moved to the edge of the bunk to see what he was doing. I watched him taking his kit out of his Bergen and repacking it again. He looked up at me and whispered, 'Is that 60lbs?' How would I know? I saw him pack it before he went to bed last night and I could only assume he did it right. Why would it have changed? The guy was clearly jumpy, the nerves getting to him, and what I did know was that he wasn't getting the sleep he needed to give his best tomorrow. I told him to shut up and go to bed. From then on, he was 'Bell-end-amy' in my head.

The next morning I got up, grabbed my kit and got loaded with the others onto the 2-ton vehicles that bussed us down to the start – a pub called the Storey Arms, now a youth hostel. It was still dark and freezing cold as our DS moved along the line, weighing every single pack individually. Bellamy's pack came in under 60lbs so the DS found a generous-sized rock to add to it. He even signed it to make sure that the same rock finished the march as started it. As we shuffled forward to the start, I felt bad for Bell-end-amy, because now he was going to be carrying 70lbs, and I suspected that wouldn't help him.

The DS checked his watch. He was the only one allowed a watch. The rest of us didn't need one because all we had to do was follow him. 'You'll be with me,' he said, calm and clear. 'Stay with me because the group that finishes with me will pass. If you're not with me, you will fail.' His gaze took us all in. You could have heard a pin drop as his words sunk in. My heart was pumping hard in anticipation.

And then off he went. Forty of us stood frozen for a second until it sank in. *Fucking hell! What the fuck?* He was gone, almost sprinting up the hill. One by one, our brains kicked into action and then our bodies followed. We started running up the hill after him, jostling for position, elbows out, weapons banging into each other, every man for himself. I needed to get up to the front where I had eyes on the DS and just stay there. Just do not let him go. Get to the front and stay on his ass. I ran as fast as I could, thinking about how they tell you that the march is supposed to be 4km an hour, but this was way faster than that. We were running. Uphill. With 60lbs on our backs. The Fan Dance was well and truly on.

As I say, the whole point of having the Fan Dance on day two is to whittle down the numbers. There are only six or eight DS, which makes 200-odd people an impossible number to manage. I know now as a former DS that they need half of those guys to fuck off, just to make the rest of it manageable. The Fan Dance goes up and over the peak of the Fan at 2,900ft, loops back around the reservoir below and then begins a long, steady climb along the old Roman road. For the first 6km going up there, there was still quite a big group around me, but, by 10km in, it was just a small clump. Near the top, I got a chance to look back, and I could see some of the lads from our group still way behind. We were coming back over the mountain, but I could see a load of them still going down to the turnaround point. I looked at them for a second and realised they had no chance of catching us on the downhill. I thought to myself, *You might as well just give up now.*

At each checkpoint, with lungs bursting, I stopped to utter my name. The mist was coming in and the freezing rain was driving in my face. Visibility was getting worse and everyone had to check in to make sure nobody had wandered off the mountain by mistake. The DS didn't stop, though. Over the top of the Fan he seemed to hit another gear and just kept going until he was lost in the mist. There were six or seven of us in the first group, queuing up to give our names, all the while trying to keep an eye on where the fuck he'd gone.

'Billingham?'

'Staff.'

My name was ticked off and I was off again, sprinting frantically straight down the hill after the DS, my thighs and knees taking a hammering from the 60lb pack on my back. I finished 20 metres behind him, in the first group, in just under three hours and thirty minutes, hopeful that it was fast enough to pass. I felt amazing, but nobody at the finishing line was congratulating anyone. Respect? Yes. But in our minds we were still competing with each other. The DS started ticking off our names while we were looking back up the hill to see how the rest of our group were doing. I wanted to know how my mates were getting on. From the guys I arrived with, four made it. The paras always do well at selection and have long formed the backbone of the Regiment. Depending on our finishing times, we were each loaded back onto one of the buses. Everyone on my bus was pumped, confident, relieved that the first major task was accomplished. It was a strange mixture of emotions, because we all knew how hard it had been but also that this was only day two. We had another month here and none of us suspected it was going to get any easier.

Finally, the bus started rolling and we arrived back at the dorm to unpack and rest, but what we saw there was a shock. Nearly half of the bunks had been cleared out, including the one below mine, their owners gone back to wherever they came from. It was weird, almost like a concentration camp. Later I discovered that, of the 183 men who had started the Fan, only 103 made the time. Those who remained were called to the parade ground, where we stood in line again, aware that half the group had now gone. We were looking at each other with that eye of competition because we all knew that anyone who was left had a shot. I was thinking that if I was fit, then these guys were at least as fit as me. I knew they weren't going to take all of us, so these were the men I had to beat. Starting tomorrow, it was game, fucking, on!

The number of billets gradually reduced over the following weeks as more people dropped out and those who survived were subsequently put into dorms together. We began to develop a real camaraderie, while at the same time remaining suspicious of one another. Despite everything we'd heard about the SAS accepting as many people as passed the selection process, none of us really believed it. Those who had passed the Fan Dance were boosted in confidence, and perversely at the same time I think we also missed the stragglers who had made us look good. They took the pressure off the rest of us while they were there, but now everyone who was left had proved themselves to be competent and a force to be taken seriously. We all wanted to make everyone around us look as bad as possible compared to ourselves. It was a confusing time and you had to watch your back. It was all about performance.

Every day was a physical day. Every day was a test. At first, we were grouped in teams of four, then two, and then we were on our own, schooled in advanced navigation and signals, and safety on the mountain techniques, and then our DS would send us out on the hills to complete the day's task. The SAS moves at 4km per hour on every march, known as TABs (Tactical Advance to Battle). Every time I went from A to B, it was expected that I moved at least at that pace. The mindfuck of it all was that you never knew how far or for how long you were expected to keep at that pace for. Every day was hard. I'd be given a starting point and a bearing and then sent out solo, always solo. I might have an hour or three hours or four hours to make it to the next checkpoint, which was somewhere along that bearing, where I would then receive the next bearing and so on. Some days I could have 20km to cover, others maybe 50km, I was never sure, which was the point. They wanted you to always be on your toes and to be able to problem-solve at a second's notice.

The SAS are looking for mental strength and independent motivation as well as physical fitness. That's something a lot of people don't understand. The guys who make it through selection aren't always the massive He-Men with muscles bulging out of their eyebrows; a lot of the time it's the last person you'd expect who makes it through. I knew that we were allowed to fuck up twice, so missing a time on a TAB by a couple of minutes was forgivable only two times. After that, you were out on your ear. But most of the time people dropped out themselves, either because they couldn't keep going mentally with the sheer relentlessness of it, or they picked up an injury along the way that determined they would ultimately fail.

I was glad that I hadn't come the previous summer, grateful that I hadn't had to endure the heat, and actually learned to love surviving and working in the cold – except for the mist and the fog, which always made navigation a million times harder. But when I had a clear day and I was sure of my bearing and where I was heading, maybe could even see the man in front of me in the distance, I was happy and confident. I began to thrive in the face of the challenge, the sight of someone else was something to aim for, a reason to get my head down and set myself the target of catching him before the next checkpoint – I was in the fucking zone.

I passed all the TABs with no warnings or red flags. The challenges were getting harder all the time, starting to fall at night-time or to involve carrying heavier loads over longer distances or in faster times. All the time, the DS were watching me to see what I was capable of. Always pushing you further but never shouting in your ear. There was a heavy-load carry – 10km uphill with 120lbs on your back. There was the Endurance – 60km through the night carrying 55lbs, not including food, water or rifle. Every one of them was relentless, brutal and designed to find your weaknesses and challenge you to confront them. If ever I was behind the time when I reached a checkpoint, the DS would calmly say, 'You are behind the time, you need to move yourself' – that's it. It was simply down to you. And if anyone ever felt like saying, 'Oh fuck it, I don't want to do that', the DS would help to make it a reality: 'Okay, no worries. Leave your Bergen there and we'll get you off the mountain, you can go home.' There was never any external pressure. How you navigated the routes was up to you, too. Think you've got the stamina to go straight over the top of the mountain? Go for it. Prefer to navigate around the contours? That's fine, too. It was all designed to make you aware that, at the end of the day, you were your own boss.

The TAB we had over the Elan Valley in mid-Wales runs between the Elan and Claerwen rivers, along a route that required us to cross one of the tributaries of the river Elan. Where I started off that morning, the mist was right down, visibility was 2 metres at best – I couldn't even see the DS standing in front of me talking – and it was fucking cold, so I set off hard, running to get warm while frantically trying not to twist an ankle or knee as I stumbled down. I reached the river after around twenty minutes in and realised from the map that I had a choice – I could either cross it or navigate around it.

I stopped on the bank and looked closely at the map as the sweat dripped off my nose and I tried to ignore my body aching all over. I could see the bridge over the river was another kilometre downriver, which meant 2km by the time I made it back up the other side. I decided I couldn't be doing with that – I'd cross directly instead. The river looked pretty fast but I reckoned it was only waist-deep and I could run off the damp on the other side.

I lifted my rifle over my head and stepped out into the freezing water. It was so cold it took my breath away for a second but, as I started to get used to it, I picked a route across. About halfway, I lost my footing on a rock and slipped, dropping my shoulder slightly but enough for my Bergen to catch the flowing water, which took a hold of it and dragged my whole body underneath. Next thing I knew, a torrent of water was crashing over the top of me and running down my top, almost drowning me, and, because it was so cold – freezing, ice-cold – it completely took my senses away from me and I started to panic, just fighting to get my head back out of the water. It had all happened in a split second and my whole world was turned upside down.

I finally pulled myself up and righted myself, fighting the current to get back on my feet. Then I realised the worst thing I could possibly have done had happened – I'd dropped my rifle, the worst thing a soldier can do. I'd dropped my fucking rifle in the fucking river. *NO!* I got my head together fast. It couldn't have gone far so I crouched down and started to feel around for it on the riverbed, but it wasn't there. It must have been taken by the current. *Fuck!* The only bonus was the rising panic pushed the intense cold to the back of my mind.

I stood up again, trying to catch my breath, and screamed, 'Shit!' Then I put my hands back down by my feet again, but nothing. I knew I hadn't moved far but the river was flowing fast. How far could it have gone? I started shuffling in the direction of the flow, feeling with my feet and hands, for five, maybe ten minutes, the whole time panicking and trying to imagine the humiliation and embarrassment of having to go back to the start of the TAB and tell them that I'd lost my rifle. That would be the end of it – losing a rifle is the end of selection for you, no excuses. There was no fucking way on God's earth I could do that. I would rather die in that river trying to find the bloody thing than go back to the DS without it.

I searched for what seemed like for ever until I felt something under my feet. It had only moved about 3ft to my left. I'd found it. I'd been handed a lifeline. My numb fingers grapsed the butt of the rifle and I pulled it close to my chest as if I'd found the most precious fucking thing on earth. At the time, it was I suppose. I got out of the water on the far bank and took stock of my situation. I was now soaked head to foot and my Bergen weighed another 20lbs because it too was wet through. I was keenly aware I was wasting time that I didn't have. I thought to myself, *Fuck it*, and, keeping a firm grip of the rifle, just started to sprint for my life. As if the hounds of hell were after me. I didn't know how far it was to the next checkpoint, it might have been 7km away or 10, I didn't care. I just knew I was running all the way to keep warm and win back the seconds lost in the river.

As I ran, I started to make up the lost time and I also fooled myself into thinking I was starting to feel warm again, even though, really, I wasn't warm at all, because my clothes were still soaked in icy-cold water. Still, I was catching and overtaking other people along the way, because in my mind I couldn't afford to slow down until I'd reached the end. At every checkpoint, I prayed it would be the last, but it never was, just another point along the way. The Elan Valley is really open and the mist had come right down, making it nearly impossible to navigate and bringing more cold with it, too. My sodden fatigues were now an icy shroud. Eventually the TAB took us up and over the top of a ridge from where

I could see down the other side. In the gloom, I could just make out one of the big 4-ton army vehicles there. It was rare to be able to glimpse a checkpoint to come, but I needed it so badly. Thank God, it was only around 1.5km away, and all downhill.

I set off again. I was feeling fine; in fact, despite the circumstances, I actually felt pretty good, but I only got around another 100 yards down the hill and suddenly I felt it – *Wham!* It was like Mike Tyson had smacked me with a right hook. I was reeling, feet going from under me, and I actually wondered if I'd had suffered a stroke. One of the guys I'd overtaken, Andy, stopped and asked me if I was okay, which I clearly wasn't. He helped me to sit down and I told him I'd be fine; he had his own time to make so I knew he couldn't hang around looking after me.

'Go, mate, fucking go!' I shouted at him and he took off.

Those pictures we all remember of marathon runners who hit the 'wall' – well, that was me. I stumbled in a daze for what felt like the next hour, but it could only have been about fifteen minutes. I can't remember clearly but somehow I must have made it to the checkpoint next to the 4-tonner even though there was no way I could take another step. I knew I was done, my selection was over for this year; I simply didn't have enough in the tank to take on another leg.

The DS at the checkpoint looked at me.

'Are you okay?'

'I really don't know,' I said.

'Right, well get on the vehicle.'

I was gutted. To have made it that far and then getting pulled off for being so stupid, making a bad call, slipping in the river, dropping my rifle, whatever I put it down to, seemed like such hard luck after all the work I'd put in.

I slowly climbed up onto the 4-tonner. The effort was ridiculous and I stopped for a second, but my body was beat up and worn out. There were three lads sitting on it. Andy was one of them. Surely they hadn't all been pulled off? It slowly dawned on me that it must have been the last checkpoint. I hadn't been pulled off it at all – I'd passed it. I'd passed it by the skin of my teeth, but I'd passed it. I'd made it through the Elan Valley TAB and survived to fight another day. It scared the shit out of me. I put my head between my knees and stared at the floor. That night when I got back, I ate as much as I could, took on as much water as I could and went straight into bed, where I lay awake for a while going over where I had gone wrong on the course. *That was too close, Billy,* I bollocked myself. *That's how people die in the mountains, because you think you're all right when you're really not.* I still didn't feel great, but I knew I'd pushed my body right to the edge of where it could go; I'd made myself vulnerable and that really fucking scared me.

—

Six hours later, I was back on the parade ground for the next day's challenge.

Selection is testing more than your ability to navigate, your endurance, your stamina and physical fitness.

What they're really looking for is the will to keep going. That's why they're not screaming at you; also why they're not encouraging you either. They're just asking you, 'Do you have that will?'

It was the first time in the military that anyone had *asked* me to do anything. A lot of people can't get their head round that. Everyone expects the shouting and screaming like they've seen on the TV, but, in reality, it's nothing like that. The second you want to go home, they say, 'Okay, take your kit off, lay it down there, someone will take care of that for you. You just make your way back to barracks and pack up your stuff.' They couldn't be nicer about it, as if you were checking out of a hotel.

But I knew to my core I wasn't going to quit. I was starting to feel tired by the third week. I had twisted my right knee and there was a lump of fluid beginning to build up on the back of it, but it kept me going. I knew that I had only a few more days until the Endurance, the last march in the hills, and if I made it that far, I just had to get through it.

The Endurance is as bloody tough as it's name suggests. It's a 60km march through the night, carrying 55lbs of weight. My knee was so swollen that night that I could hardly stand up, but I took a mouthful of painkillers and thought to myself, *Fuck it, it's another sixteen hours to go, and then I'm there.* By the time we reached the Endurance, there were thirty-five of us lined up out of the original 183. As we prepared to start, it dawned on me that I'd already achieved so much just to still be there. A lot of really strong, fit, determined guys had dropped out already and there was no shame in making it this far. That was a huge confidence boost and I knew every man on that starting line was feeling it, too. The mood among us all was brilliant; we'd bonded and the tensions and rivalries that were there at the start had ebbed away, and now we were even having a good craic with each other. The paras were taking the piss out of the marines and the marines were giving it back as usual, an army rivalry going back longer than anyone can remember, but we were all friends now.

But none of my new friends were going to help me with the final task in hand. Endurance is something you have to conquer alone. I knew that I had to finish this, whatever anyone else did. The only thing in my mind was that I had to finish, had to pass this last test.

Sixteen hours later, after a lot of blood, sweat, swearing and cajoling, I crossed the last checkpoint. I was so fucking close! I had my name ticked off and loaded up onto the buses with other lads. I was spent, tired, broken. I'd given it everything I had and, whatever the outcome now, I knew, just like when I finished P Company, I wasn't ever going to do it again. It was do or die.

We got a couple of hours to rest, shower and clean ourselves up and then the last thirty-five men assembled in the cookhouse to hear whether we'd passed the hills stage or not. I sat at the long trestle table, clutching my warm cup of tea in both hands. I could hear the other names being called out, followed by the words, 'Passed, Passed, Failed, Failed'.

I couldn't concentrate on any of them long enough to hear who had done what, I just focused hard on my own name until finally it came.

'Billingham.'

'Staff.'

'Passed.'

'Staff.'

The greatest professional achievement of my life so far. I'd passed the hills. I was through to the next stage of SAS selection and, in a way, I didn't care if I passed or failed that, because I'd proved myself here and now. I had credibility. All I could think was, *Fucking hell, I've done it, this is me now, this is happening.* I could hold my head high and nobody could take that away from me.

I was one of twenty-five men who had that feeling that morning as we filed out of the cookhouse to have our blood tests done by the doctor, to make sure no one was on any performance enhancers. Two of my toenails had dropped off, my feet were covered in blisters and my knee was swollen up like a balloon, so the doctor gave me some more painkillers. Then we loaded up onto the buses for the last time, to return home. I had two days to rest and see my family, for the first time in a month, before I would be back for the next stage of selection.

I had no idea what was about to come.

—

I enjoyed a much-needed break over the weekend with Julie and the kids, where I could recharge my body and brain – even if only for forty-eight hours. I had immediately fallen into bed when I got home, and then sat down at the dining room table and eaten for Britain. The hills had taken more out of me than I thought possible, but I was buoyed by the fact I had completed it. I had completed it! No one could take that away from me, and I was now in contention to take the next phases by the scruff of the neck and actually pass the whole bloody process – to gain entry to the Special Air Service: what I'd dreamed of doing, but did not dare believe could actually happen. Some seriously good blokes had failed the hills, yet here I was, anxious to crack on with the second phase. I headed back to Hereford to begin two weeks of training before I was due to fly out with the remaining guys. I was feeling pretty confident.

A lot of the details of the SAS selection process are shrouded in mystery, but, having already completed a tour and training in the jungles of Belize with the paras, I felt like whatever was coming next would be relatively easy for me.

Like all the lads, I was getting more and more excited following the hills, because there was a growing feeling that we were almost there. Personally, I thought, *Well, I've done the hills now, so the rest's going to be a piece of piss.* I had no fears about what was coming next and, in fact, I was just enjoying all the preparation and training. As the next phase approached, I was completely relaxed about the whole thing, thinking to myself that this was going to be a walk in the park. How wrong I fucking was! This would test me beyond anything I had yet encountered, both physically and mentally, and I needed all my skills, toughness and stubbornness to complete it.

It's no secret that I made selection to the SAS, nor that I served twenty-plus years with the world's greatest elite force. But, when I signed up to serve with the Regiment, I did agree to maintain confidentiality as to my service with them, and I am bound to honour that agreement. My years spent serving in the Special Air Service were varied, exciting and often highly dangerous, across many arenas of conflict, and I have drawn from that experience in the course of the stories I have told in this book, while ensuring that I continue to protect national security, ongoing operations and the lives of others.

As one would expect, the Regiment served in areas where people can die or be seriously wounded – our own people, as well as civilians and the enemy. This book isn't a story of glorifying conflict. But, for the record, we served with honour, integrity and bravery – always. I was often in situations few will ever have to face, but our training enabled us to survive and to protect life. I lost a few close friends, and I think about them all the time. I also made friends for life, who I treasure. The fellowship of serving in conflict, where every decision is a life-or-death one, forges bonds for life.

It was a privilege to serve. But, like everything in life, nothing lasts for ever.

CHAPTER 7

Hollywood Calling . . . Standby

Every warzone had a massive impact on me. Every time I was back home, I felt that I didn't have enough time to adjust and adapt to the change of environment, because everything seemed to move so fast with the Regiment. I was constantly on edge. I would find that even the most trivial thing could trigger a bad mood or a bad reaction. Julie might say to me, 'Oh, the dishwasher's leaking.' And I'd explode, 'The fucking dishwasher's leaking? What the fuck are you telling me about that for? I've just been in a fucking place where people are being blown up and you want to talk about the dishes?'

I can't imagine now how Julie must have felt. While I was away, she was back home looking after our children, waiting there night in, night out for me to return, and yet I couldn't find any compassion for that inside of me. Instead, my attitude was, 'Where do you think I've just been? We have more now. You've got this, you've got that. What more can I do?'

I wasn't alone in feeling like that, because a lot of the guys who came back told me they felt the same way. Soldiering is a very unconventional way of life and a lot of us end up living two or three different lives, not just in terms of multiple relationships, which a lot of guys did end up getting into, but also in the way you act, the way you are. Sometimes it felt harder to handle life back home than life on operations, so guys would volunteer to go back out before they really had to, just looking for an excuse to go away and stay away. It was easier to be selfish.

I binged everything. I binged sleep. I binged fitness. I binged drinking. I binged being a father. I binged being a husband. Everything felt so short and sharp and sporadic, that life was happening so fast all the time. It was hard to settle into anything, until I became a person I hardly knew any more – almost like a Jekyll and Hyde.

Looking back, I feel like Julie had a harder job than I did. My world was full-on, everything was new and exciting and dangerous, so it took up all my time and energy and concentration. If there're any pats on the back or medals to give out then it's the women who should get them. They're the ones who were looking after the kids and trying to hold our families together.

The wives and the girlfriends back in Hereford had to keep the wheels on the bike and they had to do all that alone, while never really knowing for sure what the fuck we were doing. Yeah, Julie knew that I was abroad on operations, but she could never be sure exactly where or doing what. She must have been constantly hearing the news on the TV or the radio and dreading the worst, wondering, 'Is that Billy? Is Billy involved in that? Are my kids going to still have a dad when this is over?' Every night she went to bed dreading that knock on the door. And it happens. It happens a lot.

We're the worst. We're the shipwrecks. We're all the same, all a bunch of adrenaline junkies. I'm surprised she tolerated as much as she did. Because there was no comfort there for her. If I'd come back with a massive pay cheque, that might have been some consolation, but I didn't. I was high and dry half the time, even coming back from war. Soldiers are taxed even when fighting a war. That was something that disgusted me about the British Army: the wages were shit and, even as a sergeant major in the SAS, I was earning about two and a half grand a month. There was one time I got back with so little money in the bank that a Regiment mate, Frank, literally had to give me the last 60 quid in his wallet to help me out.

Of course, we didn't do it for the money. But not having enough made life even harder. Every time I came back from conflict, within a day or two I was looking for work on the side so that I could make some extra money. I started moonlighting, doing a little bodyguarding and security work. I thought I was doing the right thing, because we needed more money; we needed it to buy better furniture or take the kids out, so I was always thinking, *I'll get this done and make some more money then she'll be better off when I'm off again to war.*

There were periods when I probably could have spent more time at home, but I'd always volunteer to go back on serving. Whether it was an active operation or to get in on a qualification, I was always scared to miss out on something. I didn't want to hear about someone else in the Regiment being involved in a wonderful, successful mission, rescuing hostages or taking out terrorists, knowing that I could've been there but I chose to be walking around the park with the kids.

If I could change it now, I would. Of course, I realise that I'd rather have said, 'Fuck that. I'm going to be with the kids.' But at the time, I was still a kid myself, a kid who still desperately wanted to be somebody. I was still that lad from Walsall who wanted to make something of himself, to be part of a team. I wanted it, I wanted it so bad. That's a regret I have to carry, because I wish I'd been a little bit more compassionate, a little bit more understanding of the people around me, the people I loved and who loved me.

Now, I look at my mum and dad's generation and the generation before them and I admire how they all stuck together. My parents went through thick and thin together and made it work. They didn't go through what we did – our lives were particularly extreme – but still I wonder if I could have worked harder or if there was another way we could have worked things out to stay together. It would have been nice to have been able to do that. But I simply didn't have it in me to make that happen. I want to be in a position very soon to be able to become a great husband, build our family as a strong unit and be able to spend time together, do great things and learn valuable lessons. I very much want to be a brilliant grandfather, too, because I wasn't a good father. It's not because I didn't want to be but because I couldn't be.

The truth is that I loved the fear and the adrenaline mixed together. I was addicted to the excitement. I craved the camaraderie and the constant challenges. War was a buzz that I couldn't recreate anywhere else, no matter what I did. One minute I was jumping out of planes and helicopters, bursting through doors and windows, every second of sleeping, fighting, eating, moving, whatever I was doing, constantly on edge, never knowing when it was going to be finished – the next I was back at home wondering why the dishes hadn't been done.

Gradually I started to see the effect it was having on the kids and that our lack of communication made it too hard, too destructive to continue. I'd come back and I'd try to sit at the table, doing the happy family thing, but I could sense that the kids were looking at me first, then looking at Julie, to see what mood I was going to be in. Eventually, I thought, *Fuck it, I can't keep doing this.* Was it any surprise we eventually decided to call it a day and divorce? Not really. By the time I was returning from yet another classified operation in the late 1990s, it was obvious we couldn't go on. We tried our best to remain in contact for the sake of our daughters.

At the end of the day, I can only blame myself, but there are ways now, things that we should learn to do in order to help soldiers not to repeat our mistakes. Soldiers need time and help to decompress when they're coming out of conflict. I think we should be doing more of that. We never got any help. We came home from war where somebody had been injured or even killed, thinking, wondering, *Could I have done something differently to help avoid that?* We never got a chance to clear the air or to cool off over a few days, a few beers, to talk it through in depth. It was fast and furious and often I'd be on to the next job before I'd had time to shake off the last. Eventually those bits of doubt and sadness and anger eat away at you, floating around in your head. And then it's no surprise that you end up snapping and taking it out on your family or volunteering to go and take on some extra cash-in-hand work. My relationships always seemed to get sidelined due to my active service. In Africa, where I did a lot of training, I met Sue in Harare, Zimbabwe, and we were off and on, but still managed to have a son (Jake) and a daughter (Mia) by the turn of the millennium. But how can anyone maintain healthy relationships with the life I was leading? The world seemed to go into free fall once 9/11 brought the destruction of the Twin Towers and the USA and Britain finally decided to end the reign of Saddam. It was a campaign that I played a part in. I cannot and will not discuss this part of my service for obvious security reasons.

After the Iraq War, I got a call from a friend, Tony, who'd been in the Regiment with me, asking me if I could do a quick security job – a week looking after a high-profile celebrity and his girlfriend at a film festival. The money sounded great and I could take a week's leave no problem to fit it in.

'Sure,' I said, 'why not? I could do with a break from home. Who's the celebrity, by the way?' I'd almost forgotten to ask.

'Tom Cruise.'

There's a well-trodden path between the Regiment and security, and a lot of lads end up doing it after they retire from the SAS. I think that's maybe because so much of what we do in the Regiment is security. Security was our bread and butter and I'd done a lot of it during my years of service days, especially with protecting high-profile dignitaries. I was skilled in making sure these individuals could go about their business without getting kidnapped or worse. We adapted the techniques we learned from our specialised employment within the Army. We modified our tactics to keep film stars safe from paparazzi and their crazed fans.

A lot of the guys in the paras and in the Regiment got involved in moonlighting to earn extra cash. It was almost a rite of passage that, as your career developed, your family grew and you needed more money, there was extra work that you could do. At one point, I was driving a truck on the weekends for a construction company building the new stand at Aldershot football stadium; another time I was doing security at music concerts. I did what I had to do to make the extra cash that we needed, and we needed it all the time.

Tony's company was probably the top one and did a lot of bodyguarding for the most famous celebrities. One of his guys had let him down at the eleventh hour, so I was really helping him out by jumping on a plane over to Rome for four days to look after Tom Cruise, otherwise the contract he had might be put in jeopardy. Up until then, I'd done a few bits and bobs for Tony, helping out as part of a team for a day here and there. I'd worked a Bryan Adams concert and done a couple of days looking after Kate Moss during her daughter's christening, but this was a bigger gig – four days working alone with the client over a long weekend, which meant good money and a chance to prove to Tony what I was capable of.

I really felt like the risk of losing my job was worth it, for the potential upside of getting in with Tony and maybe even pursuing the bodyguarding work when I eventually left the army, so I felt it was the right decision at the time. I certainly wasn't disrespecting the SAS. There was also a little bit of extra excitement because of who that client was. I'm not a groupie by any means – it takes a lot to impress me – but even I'd heard of this guy. For the next four days, I would be responsible for the safety of the world's biggest movie star. Tony briefed me on what Tom liked, what he knew, where he had to be, where I was going to pick him up, what the basic itinerary was etc. I had a small bag and a change of clothes and off I went on a plane to Rome.

The driver was waiting for me at the airport with a sign with my name on it and we drove straight to the hotel where Tom and I would be staying, so I could get cracking. The first job was to recce the hotel, work out how I was going to get Tom in and out safely and with the minimum fuss.

I checked in and went up to the top floor of the hotel. Because it was Tom, we had the entire floor to ourselves. After I'd recced the whole floor – where my room was, where Tom's was, where the escape routes in and out were – I went back down to check the outside. I knew Tom was flying in from New Zealand, where he was shooting *The Last Samurai*, and he would be going straight to a restaurant to meet a film producer that night, after which we would return to the hotel. When I came back down, I could see that a bit of a crowd had started to gather outside of the front of the hotel. I slipped outside to take a closer look and my heart sank when I realised that a bunch of them were paparazzi. I sidled up to one of them and asked what was going on.

'Tom Cruise is coming,' he said with a big, happy, excited grin.

For fuck's sake. It was all supposed to be hush-hush. Nobody was supposed to know that Tom was even in town, let alone where he was staying. Now I was in the shit because I was totally on my own and there were a hundred people blocking off the entrance – the only way in and out. I only had one option – I needed to negotiate with this dude.

I introduced myself and explained that I was Tom's security. Pepo told me that he and his mates were all local paparazzi and that all they needed was a few shots of Tom.

'Look, Pepo,' I said, 'I can't bring him here. That is not how it will work.'

Pepo looked pissed off.

'There's too many people, mate,' I said. 'I need to take him to another fucking hotel unless we can work out a way to keep Tom safe and still have official shots taken. How about you help me to get all these people out the way?'

My offer was that, in return for a few seconds of Tom outside the hotel when they could all take their standard photos, Pepo promised me that he would clear a metre-wide channel for me and Tom to walk from the car to the front door. That seemed fair enough to both of us, so we agreed and I said we'd be back later. Tom flew in on a private jet and, as he came down the plane's steps, he saw me and immediately smiled. 'Billy?' I was impressed he'd done his homework. Of course, I already knew who he was. I'd managed to get the car right up onto the tarmac (it's amazing what doors the name Tom Cruise opens), so I was able to get him straight off the plane and to his meeting at the restaurant, briefing him on the way as to how the evening was going to go and what was going to happen when we got back to the hotel.

Tom was great – happy with everything that I'd set up and happy to go along with the plan, once he fully understood it. A proper pro, which made my job easier. So, right after his meeting ended, we drove back to the hotel. I was still a little nervous, because, despite what Pepo had said, crowds can always be unpredictable and you never know when a crazy fan is going to decide now is their moment to get up close and personal with their hero. My face fell as soon as the car pulled up outside the hotel, because the crowd of 100 was now 300. I was well and truly fucked. How the hell was I going to get Tom through that lot all on my own? I needed a whole team to get into that. Tom knew it, too. He gave me a look as if to say, 'What are we going to do here?'

I wasn't going to panic. I decided to see how we would manage this.

'Let me get out the vehicle, Tom. Lock the doors. Let me just do some negotiating and then I'll come and brief you on what's going to happen next.'

'Okay, Billy,' he said. The guy was so cool. He impressed me.

The crowd was huge. The noise of it hit me the second I got out of the car. All I could think was, *What the fuck am I going to do now?* I couldn't even see the front doors. This was my first solo job, with Tom fucking Cruise of all people, and it was going to go tits up. There was no back route into the hotel and it was too late to be looking for somewhere else that I could ensure was safe. I walked around to the side of the crowd, looking for Pepo. Eventually I saw him.

'Pepo, what the fuck is this? I already told you, I will not put Tom in the middle of this scrum.'

He looked at me for a second, shrugged as if to say, 'What's the problem?', and then walked over to the crowd and shouted at them in Italian. All the people just started moving. I couldn't believe it. It was like watching Moses parting the Red Sea. I looked around and there, sticking his head out the window of the car, was Tom, like he was watching a miracle. I was impressed. More importantly, Tom was impressed. I walked back to the car.

'Happy?' Tom asked.

'Yeah, let's go,' I said, and Tom hopped out of the car.

I took a second to explain exactly how it was going to work.

'Just stand on my left-hand side, Tom. I don't envision having any issues or troubles, but if I do have to grab you, please don't take it personally. I've got to do what I've got to do.'

'No, I understand. Security is what you do. You do what you have to do.' He was really cool.

'I appreciate that, Tom. That's great. Let's go.'

So we started walking through the gap in the crowd. Hundreds of people were screaming Tom's name either side of us. My heart was racing because that was the prime time – if some dick was going to jump out and try to grab him to give him a cuddle or worse, then that was the moment it was going to happen. Just before the doors I stopped and pointed Tom towards Pepo and the other photo guys.

'Can they have a few seconds for photos, Tom?'

'Sure, that's fine,' he said, smiling at the throng.

I started counting down. 'Ten, nine, eight, seven, six ... okay, done.'

I went to turn around and that's when I saw it. Tom and I were a step away from the door, when out of the corner of my eye I realised that somebody was there. Someone was moving towards Tom, fast. I went into protection mode. *Fuck!* I grabbed Tom around the shoulders with one arm and at the same time I reached out with the other arm and just grabbed this thing by the arm, too. I couldn't see a face, just jeans, T-shirt and a blue hat. I didn't have time to think, I just *did*. With Tom under one arm and the suspected attacker under the other, I pushed towards the door and we fell inside the hotel doorway. It was all reactive. I looked around to check Tom was okay but something didn't add up. Tom was smiling. That massive big, genuine smile of his was at full throttle. But he wasn't smiling at me; he was smiling at the person I had under the other arm. I turned and realised that under my other arm was a woman, a beautiful woman, who was also one of the best-known actresses in the world.

Before I could say anything, Tom piped up, 'Billy, meet Penelope!'

I'd just saved Tom Cruise from his own bloody girlfriend – Penélope Cruz.

—

My first marriage ended in divorce and my relationship with Sue ended the same way. I found myself nearing the end of my Regiment career, single, with five kids, not really knowing what the future held. It is a hard thing to accept that your time in the Regiment is coming to an end. But the truth is that everyone's time comes and everyone has to move on. Two years of being the sergeant major in Iraq was particularly exhausting and I knew before I came back from Baghdad that the time was approaching to step aside and pass that mantle on to the next bloke. My days of kicking in doors and jumping through windows were over, but there were plenty of younger guys coming after me to take that over. I was gutted, but I knew it was the right thing to do.

So, in 2006, when I was offered a posting back to the jungle, training some of the younger army recruits following after me, I jumped at the chance. I felt it would be good for me to get away and back to the familiar. I had missed being in the jungle and, professionally, I needed something that could fill the massive hole that being sergeant major in the Regiment had left. I knew I could never replace that role, it was my dream job, but the next best thing was being in the jungle, the place I loved, training people, the thing I loved doing. The army had established a huge training facility, providing A-grade instruction to all areas of the army as well as to Americans, Canadians, New Zealanders, people from all over. The role I took was the top job and I was really looking forward to getting stuck in. I knew it was a tough course and it would be hard work but it was just what I needed. The position meant that I had several DS operating under me and the course itself was instructing and teaching people, rather than just selecting. We ran a jungle warfare course, where the recruits learned how to operate and fight in a jungle environment, and a tracking course, where we taught them how to follow people covertly in the forest like a real-life Tonto. In all, I spent seven of the first twelve months under the canopy.

It was a busy old time but also a lot of fun. The garrison back at camp was a beautiful place to come home to after sleeping out in the forest. There were always interesting people circulating through and lots to do. The camp had a bar with a thriving social scene, a football club and a swimming pool. It was there that I met Jenny, who eventually became the mother of my sixth child, Darci. I loved the job but I was still getting itchy feet, looking around for what else I could do and what I might move on to when I eventually left the army. I knew it couldn't go on for ever.

There was a job opening in the City of London and everyone was encouraging me to take it. London would be a good opportunity to meet influential people, make some contacts that could help me out when I left the army for good. It was the smart career move, so I applied for it, got it and flew back to England in 2007 to join the Honourable Artillery Company (HAC).

The HAC is a Territorial Army regiment, with a long and proud tradition, which essentially exists to provide support to the infantry. The company is made up of a lot of bankers, lawyers, some CEOs and guys who are based in London, but they do a great job, even getting thrown into conflict when they're needed. When I'd been in Iraq, the HAC had come out to relieve the troops on the ground in Basra and done a really impressive shift.

It should have been a nice easy gig for me, a welcome reward at the end of a long career. It was also a good chance for me to be back in the UK for a while and maybe see some more of my family. The problem was that the officer in charge and I had a huge personality clash. He also had a personality clash with the regular PSIs (permanent staff instructors). I think he'd have had a personality clash with his own shadow. In a dark room.

No matter what I did to try to broker the peace and get on with business, the situation reached a head again and again, until, by three months in, I was ready to smack the bloke. Everybody could sense it too, so the CO took me to one side and said, 'Billy, you need to take a break.' It was good advice. I'd been in the army for twenty-five years. I had a long and distinguished service record and I was in danger of throwing it all down the pan by beating the shit out of a commissioned officer. That was career-ending, no matter how big a prat I thought he was. You didn't pull that sort of shit without paying a hefty price for it.

The timing was pretty good because Tony had been in constant touch over the previous few months. Despite the Penélope-gate incident, Tom Cruise must have given me a glowing reference, because Tony wanted me to go and take on another security detail with another high-profile client. Actually, *clients*, plural, because these guys were a couple, and they were as famous as each other. The next day, I flew into Prague and met another guy, one of the security team, at the airport. We drove for an hour to the outskirts of the city, into a district filled with huge mansions, heavily gated and fortified – obviously the part of town where the rich and influential lived. We cleared security at the gates of a massive mansion and rolled inside to an underground car park.

I noticed them right away as we parked. He was rummaging around in a pile of suitcases, dressed in shorts, flip-flops and a vest; she was standing behind him in a pair of sweatpants, shades perched on the top of her head. I stepped out of the vehicle and they stopped what they were doing and came right over to shake my hand.

'You must be Billy,' he said, his million-dollar smile glittering under the strip lights.

'Yeah. Hi,' I said.

'I'm Brad.'

'And I'm Angie,' she said. 'Welcome to the family.'

Angelina Jolie was top billing on a new movie called *Wanted*, ironically about a professional assassin. The movie was shooting in Prague for three months.

Angie and Brad Pitt had a life that has been well documented in terms of work/life balance, which I would argue they got spot on. Their kids were well looked after, they still found time to undertake numerous events with their charity work, and still made great movies. It meant that their lives were non-stop busy and they needed 24/7 security, 365 days of the year. That night we sat together over a drink as they ran me through everything. Initially, my task was to look after the security, so they explained to me as much about them as they could. I was happy to have that job and I felt confident when I assured them that they would be safe in my hands.

The week was full-on. From the second I woke up, I was aware of attention from paparazzi, or the attention they got from crowds of people all the time; it was unbelievable. It felt surreal to me – to have been part of one of the most secretive organisations in the world for the past fifteen years and now to be thrust so centrally into the limelight just didn't feel real. It went so well that I was then asked if I could stay on and work with them for a longer period.

I'd only been given a week's leave by my CO, so I was due back at work in London on Monday morning. Going with my new job to Europe would mean I was technically

AWOL from the army, a court-martial offence. But I didn't want to go back. I'd already had enough of London and the politics and I was enjoying the work with Brad and Angie. I thought to myself, *Why not ride this train and see where it goes?* A week later, I was sitting in the garden of a beautiful European chateau. Across the table from me was Brad, while the kids were running around on the lawn. We were talking about upcoming security arrangements when my phone rang and Brad looked at me as if to say, 'Take it.'

I instantly recognised the voice I'd been arguing with back in London.

'Sergeant Major, can you be in Horse Guards for two o'clock?' he said.

'No, I can't. And don't call me on this number again,' I said and hung up.

I may have been sunning myself in a beautiful garden in Europe, thinking I was the 'man', but I was now the man whose phone was ringing again.

This time, the adjutant was on the other end of the line.

'Where are you, Billy?'

I paused for a second to think what I was going to say.

'And don't fucking lie to me,' he said, 'because I've got a copy of *Hello* magazine in front of me and your fucking mug is on the cover.'

Fuck. I was the man all right. I was the man in very deep shit.

I'd been caught in the background of a paparazzi shot from the previous week in Prague and the picture had ended up on the cover of the magazine. There was no point in me trying to lie. I had to come clean.

I'd had to accept for a while that I was getting older and that there simply weren't any operational postings for me any more. Everyone knows that in the army – the time comes when you have to step aside and let the younger blokes come through. I also felt like I'd given a lot back already, training those young lads coming behind me, first in the depot in Aldershot and then with the jungle training abroad. I'd given the army twenty-five years of my life and I didn't owe it anything.

The CO came on the line. He was pissed off too and he gave me a right bollocking, but then he asked me a simple question.

'Are you happy where you are, Billy?'

It stopped me in my tracks for a second. The truth was that I was happy. I was doing something outside of the army for the first time in my life, something that made me feel challenged and excited and that I was making an important contribution again. I didn't want to go back to London, I didn't want to go back to the army, I was done. I suddenly realised that this was it.

'I am, sir,' I said.

'Well, then, you'd better stay where you are.'

And that was it, after twenty-five years of service, I was out of the army. It was that immediate it left me momentarily stunned. I hadn't expected that reaction. It was a massive relief to hang up on that call. I couldn't believe my luck. I'd gone from almost being the first sergeant major in the Special Air Service to be charged with AWOL, to being given an opportunity to start a new career. It felt like a weight being lifted. I no longer had to deal with the dread of going back to London; from now on I was going to be a full-time bodyguard to the biggest movie stars in the world.

For a lad like me from working-class Walsall, I'd never encountered the way the stars from Hollywood lived before. The work was new, too. I was used to the military way of things being full-on and hostile, but my new role was more about stepping down the risk and planning. Security isn't about rolling around on the floor with people. If I'm in a scrap with someone, then I'm not able to be aware of the person that I'm supposed to be protecting or thinking about how to keep them safe. A good security detail, a good bodyguard, will already have thought through the potential dangers and headed them off before they ever have a chance to come to light.

That meant that for every occasion or job, I would go and recce the venues I'd be going with my client to the next day because I couldn't trust anyone else to do it. I was constantly reviewing the security tapes from the previous day, scanning the crowds for anyone who looked suspicious – it could be a known stalker, a weirdo or just someone who looked out of place, but I needed to know about them before they ever became a problem. My work was routine and long; sometimes working sixteen, seventeen hours a day.

Working these kinds of jobs was very much like being in the military in that you're away from home 'on duty' with little time for communication, and by that I mean taking the time to re-connect with the ones you love living on the other side of the world. Naturally, this 'new life' did put a strain on my relationship with my own family because I never had a second to talk to my kids, or a private space in which to do it if I did. I had to be suspicious of everyone who wasn't in the circle of trust. If the drivers weren't regular drivers I used for clients, then they weren't to be trusted. There could be no talking about anything important in front of anyone who wasn't in the regular team for whomever I would be guarding, whether it be Tom Cruise, Russell Crowe, Michael Caine, or anyone else. That was an absolute rule, because people were being paid off to let slip information that would compromise your clients' security. Sometimes, I'd set traps and deliberately make up a location or a time when my client was supposed to attend something, mention it in front of a new driver or a nanny, then turn up and see if there were paparazzi there. If there were, I knew that person had been paid off and had to be fired.

I realised in this new role that the real risks were more to my client's reputations than to their safety. Yes, there was the odd kidnap risk that I had to take very seriously, but those came up a lot less frequently than the risk that a star would step out of a car the wrong way and suffer a dress failure, or that they might inadvertently be photographed looking the worse the wear when out with friends just relaxing. The professional reputations of all the clients I looked after were paramount, and it was my job to help them to maintain their public profile of who they were and how they led their lives. Tom Cruise was exactly how you see him on talk shows – he had time for everyone and was a really engaging guy. Whereas Brad and Angelina were great parents who just happened to be film stars.

As part of my routine in these types of scenarios, I carried spare clothing for my clients, just in case they spilled a cup of coffee over themselves in the back of the car. I knew that kind of thing would ruin their day and that meant it would ruin mine, too. I also made it my business to know their blood groups, their allergies, and was prepared, all day, every day, for anything that could go wrong. Again, just as I was taught in the SAS, I was ready for any eventuality and could handle it by assessing and dealing with the problem when it arose. I enjoyed the daily challenge of being in charge of such important people like the ones I just mentioned. After the Regiment, it was just the kind of new experience that I'd been craving. I liked being back out of my comfort zone, taking on new tasks, leading a skilled team and doing a job to the best of my ability. It was the best thrill I could find outside of the battlefield, and it didn't hurt that I was earning more money than I had ever done before, too.

Living in Hollywood, I was also meeting some of the most famous people in the world, such as John Malkovich, Robert De Niro and Sean Penn. Sean was making the movie *The Tree of Life* in Texas in 2008 when we first met, and we got to know each other quite well. I instantly got on with him, not least because the guy would never have security or bodyguards. In a very modest way, Sean always believed he could look after himself, and he also liked chatting about Special Forces stuff, asking me about drills and weapons or the experiences I'd had over a beer in the bar. I didn't realise it then, but our paths would cross again further down the line. Many times over in fact.

All the famous people I encountered also had their own security, except Sean Penn, of course. Often, if there was an event going on, the bodyguards would gather in the same place. One of the funniest times for me was taking a client to a private screening of their latest film – a one-off job. I went outside onto a balcony that looked back into the bar area where the media and cast would enjoy a drink afterwards. Stood in the doorway to the balcony was a huge monster of a man who held the door open for me as I made my way out. Just as I was about to go through, he let go of it, smashing it in my face. I glared at him, thinking, *You've got to sleep sometime, pal.*

I went over and sat down. I could tell all the other bodyguards knew each other because they were in their little clique, turning away from the new boy as I walked along past them. I took a seat a little away from them all and just enjoyed having a bit of a breather. A few minutes later, my client's make-up artist came out onto the balcony. He had been around for years and knew everyone, said hello to everyone – a life-and-soul kind of guy. He spotted me in the corner and started making his way over to me, stopping to chat to a few of the other bodyguards on the way.

'Why are you sitting alone?' he asked me. 'Why aren't you sitting with the other bodyguards?'

'Oh, I'm fine here,' I said. 'No problem.'

'Okay,' he said and returned to the bar. I could see that my client had come out again now and was back in the main bar, doing their thing, circulating, having a drink. There wasn't any threat so I stayed where I was.

Eventually he gave me the thumbs-up that it was time to leave, so I got up and started for the door. Suddenly all the huge bodyguards who'd ignored me just an hour before were jumping to their feet, moving their chairs out of the way to let me through. Some of them were looking me in the eye, saying, 'Hi' and 'How's it going?' as I passed. Even the monster at the door leapt to open it for me. This time he didn't let it go.

It's a funny thing being in the SAS. A Regiment soldier can spend his whole life trying to get into it, only to then spend the rest of his career denying he's ever been in it. The secrecy around it is second to none. Or course, that works both ways. Back in the bar, the make-up guy came over to me and explained that he'd mentioned to the bodyguards outside that I was a sergeant major in B Squadron – with 22 SAS . . . just like them. Of course, all those guys outside had made up their own SAS credentials to get their lucrative bodyguarding jobs with the other celebrities inside the bar. The problem was that none of them had actually been in the SAS, so when they heard that I was the real deal, they'd all shat their pants. The big fucker on the door, I later

I was the only skinhead among the mods at school in the late 1970s.

Even at eleven years old, my destiny was sure to be the military.

Tackling another obstacle on the P Company steeplechase, 1983.

Passing-out parade where I was awarded 'champion recruit' for 497 platoon.

Fuck, that was cold!

On my tour with 3 PARA in 1984 in Belize, and carrying the American M16 rifle.

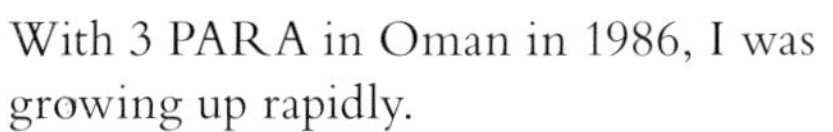

With 3 PARA in Oman in 1986, I was growing up rapidly.

Meeting the village elder in Oman on patrol with 3 PARA, 1986.

My great friend Guy Homan ('GBH') enjoying the view of the Jebel Akhdar, Oman.

Right from the word go:
'Always a little further.'

Protecting the buffer zone between Greek and Turkish communities in Cyprus, wearing the beret of the United Nations.

The Steve McQueen look always suited me.

Jungle training.

On the frontline.

Haiti. The strength of the people was unmatchable to anything I've ever seen before, or since.

George and I ten days after the Haiti earthquake had struck. This is how we lived for three months.

Meet Elton, a former street kid in Haiti, who I supported with REBUILD. He is now going to college.

The boys are back in town.

'Have you been busy, Billy?' the queen asked me, as she pinned the MBE to my chest. An incredibly proud moment for me and my parents.

Marrying the girl of my dreams earlier this year in Florida.

Since childhood, my best mate, Carl, has been by my side. He also had an amazing military career.

I always have to be on my toes when being grilled by Mark.

found out, was Italian; he couldn't even fucking spell Special Air Service.

The next morning all the cars were lined up for the stars to take them to the red-carpet gala in the centre of Cannes. My client was in car one. There was another big movie star in car two and so on. Just as we were getting into the car, my client decided that he wanted to ride with his movie-star buddy in his car, so he checked with me that it was cool from a security point of view. I couldn't see why not, but I didn't want to ride in his car on my own, so I went over to the new car with him. Who was the movie star's security guy? Only the big Italian, so I said to him, 'Out.' He got out.

I jumped into the car, trying to keep a straight face, as the boys had clearly clocked what had just happened and my client was beaming with some sort of vicarious pride that his bodyguard had trumped the bodyguard of the film star sitting next to him. Like a couple of schoolboys playing 'my dad's better than yours', he put his hand out and laid it on my shoulder.

'You're the man,' he said. I was getting the feeling he just liked saying that. We pulled up on the red carpet and, as we got out, he whispered to me, 'He wasn't SAS was he, that guy?'

'None of them were,' I told him.

He almost jumped in the air as he clapped his hands together. 'I knew it!' He had the best bodyguard for the day and that made him incredibly happy. From the time I started working with Hollywood stars I was flat out for well over two years taking on various jobs. Time off to go home and see my own family became harder and harder to get, because once I'd won a client's trust, they quite naturally wanted me there the whole time. With their hectic schedules, it meant a new movie always took priority – which is natural, and you are there to do a job as you are on the payroll. With the majority of my clients I wasn't part of their long-term set-up anyway, so it didn't cause any issues for either side. Even with Brad and Angelina, I eventually had to say goodbye as their schedules couldn't fit in with what I wanted to do – which was to be nearer home. Although it was a hard decision to make, I knew it was the right one as I wanted to see my kids, be closer to Hereford where the Regiment was based, and hopefully plan another adventure. Before any of that happened, though, it was time to accept an invitation from Her Majesty the Queen.

CHAPTER 8

Doing the Business

After I left Hollywood, I was back at home, working out where I wanted to go next with my life. I was craving the action of the war zone, so I decided it was time I became more operational again, but, before I went anywhere, I had one thing to do. I had received a letter from the Palace. The Queen's lady-in-waiting had written to me to let me know that 'my presence was requested by Her Majesty'. I was going to be decorated for my military service with an MBE. It was actually the fourth time that I'd received that letter. I'd had one a year since 2005, but every time the letter came, I'd been fully engaged in operations or overseas training and was never able to attend. I always thought that it was more important to continue what I was doing than to stop and collect an award or a medal or whatever for what I'd already done. But this time, I decided that there was no reason to delay it further. I called my mum and dad and told them that I was being decorated by the Queen and that we had to go down to Windsor Castle. They said they'd come along.

I'd never been in a castle before, but it was exactly as I'd expected the inside of a castle to look. There were big oil paintings on the walls, beautiful rugs on stone floors, suits of armour all around the place, everything looking absolutely gorgeous. What did surprise me was the size of it. The reception was set for four or five ex-military guys just like me who were getting medals, and the room that we were led into when we arrived was kind of overwhelming – it was so huge. While we tucked into the champagne and canapés, Her Majesty's lady-in-waiting came over and introduced herself. She ran us all through how the ceremony would begin, the order we'd be called in, so we had a complete run-down of how the day was going to go.

All the other lads and I were then led into another room and told to wait until our names were called. When I heard my name, I came marching in, eyes front, dressed in full uniform, straight towards the Queen, who was raised up on a little platform at the other end of the room. The lady-in-waiting read out the details on my citation, listing the various hostages I'd saved, the terrorists I'd captured and the work I'd done in London after the 7/7 bombings – all the things that I'd done in the SAS that they were allowed to talk about. As I walked and listened to all the things I'd done, I could see my mum and dad out of the corner of my eye. They were sitting to the left of the Queen, and I'm no lip-reader, but I could clearly make out my father leaning into my mother and whispering in her ear, 'He did what?!'

I had to stifle a chuckle as I saw her shrug, just as surprised as he was, and reply, 'Don't ask me. I don't know.' My poor dad couldn't believe what he was hearing. In all the years I'd served, we'd never once discussed the details of any operations that I'd been involved in. Now there he was, listening to the details of his son's whole career being read out to him and the Queen by some posh lady in a hat.

As the Queen pinned my MBE to my chest, she went to address me, but just then the lady-in-waiting piped up: 'And can I say that this is the only man who's ever kept Her Majesty waiting for four years.' There was a ripple of laughter in the room before Her Majesty quipped, 'Been busy, Billy?' I smiled and thanked Her Majesty. I looked back at my career, releasing hostages, firefighting on the battlefield, capturing war criminals or taking out terrorists, and I realised that, yes, I had been busy. I'd served my country to the best of my ability for most of my life. I had put my life on the line more than once and I was happy to have done so.

I thought about all the lads with whom I'd served over the years in foreign countries, who'd done the same but hadn't been as lucky as I had. I thought about the ones who never made it back at all and I felt proud of what we had all done. And I felt good, as well, now that I knew my mum and dad could feel proud, too.

—

While I'd still been in the SAS, I'd had a call from Frank, my old mate from the Regiment, who was struggling a little, looking for work. I'd never forgotten how he'd reached into his own pocket for me when I was having hard times of my own, and I really wanted to help him out. Another friend of mine, also called Billy, also ex-Regiment, had set up his own security company called Pilgrims and was making a real go of it out in Iraq and Afghanistan. I called Billy up and asked him if he could throw some work Frank's way, and so, as a favour to me, he did. Frank flew out to Iraq shortly after.

When I'd been in Iraq as a sergeant major, after Baghdad had fallen, I'd been an informal adviser to some of the American and British security companies who were working out there. Often, I'd be asked to recommend companies or individuals who were helping to gradually relieve some of the pressure on the military, and so I'd end up throwing jobs their way. After a while I started to think that I was probably not capitalising enough on those contacts, so when Frank had the idea that we could use his new experience in Iraq along with my contacts to set up our own thing, I thought I'd go for it. We decided to call the company Sabre, but as I was still serving and Frank was older, the deal we came up with was that Frank would run the business initially, while I remained a silent partner, and then when I eventually came out of the military it would be my turn to take over.

Frank did very well very quickly. He found another silent partner, a wealthy Iraqi who helped with some of the local knowledge as well as the finance. He bought a big slice of the equity in Sabre and, with his help, Frank very quickly started to establish a business out there. He employed a few ex-Regiment guys and built a fresh brand. Sabre was the first security company out in Iraq that operated with the same low-key approach that the Regiment would have employed. We used beat-up old vehicles and local people, and tried to stay under the radar when many of our competitors were drawing attention to themselves with brand-new Land Rovers and Oakley sunglasses and all that showy crap.

Because we were employing people from the local areas, talking to the sheiks, we were getting access to places other companies couldn't reach. We employed drivers from Fallujah or we agreed to build a well there, so we got access to Fallujah. It sounds simple but we were in places where other companies were shut out. We had a rapport with everyone. It was a good strategy and the business went from strength to strength. An official report published on security contractors in Iraq from 2003 to 2008 shows Sabre up there in sixth place, with a turnover of nearly $300 million.

From security, we began to branch out, finding new opportunities to build military-style camps for companies operating out there. We started selling and hiring equipment that we had access to. We were up to all sorts. Frank checked in with me regularly, reassuring me that once he'd made his millions, he would be ready to step aside and let me have a go at the helm to make mine. In the meantime, I was happy in the Regiment, giving advice from the sidelines. It seemed like a great set-up for us both.

By the time I'd had enough of bodyguarding the celebrities, I was missing the action zone a little, and Sabre was turning over millions. Frank needed more support on the ground too, so it seemed like a no-brainer that I'd go back out to Iraq and get more directly involved in the business. It would be a good way for me to learn the ropes a little better too, so that once Frank stepped aside, I'd be ready to take over.

One of the main things we were looking at back then was our staffing costs, because we were employing over 3,000 people. Security is a cut-throat business and, to stay competitive, you have to be able to provide manpower at a lower cost than your competitors. Long gone were the days of paying British guys a couple of grand a day to do basic contractor work; instead, we were looking to places where we could find men for a fraction of that.

Initially, we had the idea of recruiting Ugandans. We figured that they were all ex-British colony, they spoke English, and we could get them to work for a lot less than what we were paying Europeans. So I spearheaded a programme to fly out to Kampala to secure employment contracts with the government that would give us permission to train up to 1,000 of their ex-military guys and then fly them to Iraq to work for Sabre. We'd provide food and accommodation, internet, social expenses and pay them each $600 per month. They were thrilled.

But the bottom line kept putting pressure on us to get our costs lower and lower, so after Uganda we set up next in Kenya, where we did exactly the same but, this time, the guys would be signing up for $400 per month. It was vital we got governments on board because we didn't want anyone to think we were mercenaries or trying to recruit an African army to go and fight ISIS or any shit like that. In Kenya, that meant getting the labour minister on board, but when I met with him he wanted a bribe. He flat-out passed me an envelope across the table and said, 'What's in this for me?'

'I'll tell you what's in it for you,' I said, 'the chance to say that you've created a thousand well-paid jobs for the Kenyan economy.'

'No.' He slammed his hand down on the table. 'No.'

I stormed out. I don't do bribes. I wasn't going to get involved in that sort of corruption bullshit; I'd rather we just pulled out of Kenya and found somewhere else, so I went back to my hotel and prepared to leave the next day. While I was in the bar that night, a big lad approached me. He knew my name.

'Billy, come and talk to me outside,' he said. I looked outside. It was dark. Fuck that.

'No way,' I said.

But there was something about him that wasn't that threatening, so I reluctantly went against my instincts and followed him out.

'Please, tell me what is going on,' he said.

So I did. I explained about the meeting with the minister and why I wasn't going to play that game. I said that the offer was still on the table but only as long as I was in the country, which meant it expired tomorrow. He thanked me and said goodnight.

The next day, I got a call from our local fixer. He had important news – the minister I'd met the day before had been shot. He'd had a nasty accident with a bullet and I was asked if I'd hang around for a few days. The deal, it seemed, was back on. More contracts came up but, again, we had to cut costs, so we had the idea to try Sierra Leone, where the cost of living was even lower. I made contact with a guy out there, called John, who said he'd meet me at the airport.

Anyone who's flown into Freetown will tell you that you have to cross a body of water on a boat right after you land, but before I even made it that far, I was met at the airport by John, carrying a board with my name written on it. Standing behind him was a crowd of fifteen, maybe twenty people, all wearing T-shirts with my name emblazoned across their chests. John was short but 3ft wide, which only exaggerated that he was flanked by two bodybuilder types, 6ft 6 square, either side of him.

'Mr Billingham?' he said.

'John?' I hoped.

He gave me a massive smile and hugged me. 'These are your bodyguards,' he said. 'And these are your people.'

Just then the fifteen-odd people in Mark Billingham T-shirts burst into song. Like a gospel choir, they started singing my name over and over. 'Welcome, Mr Billingham, Mr Billingham, welcome, Mr Billingham, Mr Billingham.'

I wanted to pinch myself. What the fuck was this? I followed John outside only to find another bunch of them, also dressed in Billingham T-shirts and joining in with the singing. There was a massive banner behind them too, with my name written in giant letters across it. It was madness. I was only here for a recce; I had no idea if we could even make this thing work.

'These are your people,' said John, now pointing to a group of 4 × 4s containing a whole bodyguarding team. A crowd of people flanked us all the way down to the boat.

John started explaining how the minister was on his way and how, though the president was out of town now, he was looking forward to meeting me later. I was thinking, *Fucking hell, the president? This is all a bit much. What's he promised them?* I was actually a bit worried about what I'd gotten myself into. We got to the ferry to cross into town and I was delighted to get on board, away from my new fan club. I was trying to work out if I'd been stitched up. The Regiment had been involved in Sierra Leone a few years before, in an operation that had seen action. People had died in Operation Barras and I was wondering whether someone wanted revenge for it and I was a target. Just then, two dwarves appeared and started doing some spooky voodoo dancing for me. I was sure I'd fucking lost it then.

I called back to the office to talk to Frank. He just started pissing himself laughing at me. 'Keep safe, mate,' he said, 'let us know if you need anything.'

The ferry reached the other side and things only got more intense. The crowd was even bigger than the first welcoming party had been, over a hundred of them, and they all knew the 'Welcome, Mr Billingham' song, too. We got into another 4 × 4 and John started opening the sunroof.

'You must address the people,' said John, gesturing that I should stand up and talk to the crowd. My knees were actually shaking a bit. The crowd were still in full voice, singing the song and dancing around like it was Christmas.

'Thanks for coming down,' I spluttered. All I could see were the whites of a hundred people's eyes looking back at me. 'Thanks for making me feel so welcome.' They all started cheering and going mad again as I got back in the car and we drove to the hotel.

The following day, we had a meeting at the government building. The labour minister was there, the armed forces minister, the minister for police, all the bigwigs lined up along the top table next to me. I stood up to make the presentation I'd prepared, which I was a little nervous about, to be honest, because Frank and I had discussed it before I'd left and he'd insisted that $250 per month was the highest we could go. I was really embarrassed that that was going to be our offer to these guys and, even though I was happy with telling them what the job would entail, the leave, the two-year contract and the training they'd get, all the good stuff, when it came to the salary part, I was almost holding my breath as I said it.

There was a sort of gasp in the room. I knew it. We'd undersold it; there was no way these people were going to work for that. I finished the presentation and was fucking delighted when we took a ten-minute break and I could escape outside for some air. The minister for the armed forces came running out to see me.

'Can we talk?' he asked.

'Sure,' I said.

'Please,' he said, 'this $250 is a problem.'

I was about to start apologising, explaining that we couldn't really afford to go any higher, margins were tight, bottom lines squeezed etc., when he interrupted me.

'Some of the generals are saying they will resign their commissions so that they can join your company,' he said. '$250 per month is much more than even they earn.'

I thought, *Fuck, they were happy with the offer.* To me, it seemed like a pittance but, because the standard of living out there was so much lower, the price of everything so much less, it was the equivalent of a brilliant salary. They weren't pissed off at all. Except this guy, who was now panicking that the whole national army was going to resign and apply for a job at Sabre.

'Please, go back and explain that only retired soldiers and police can apply,' he begged me.

—

The business continued to grow. We were winning contracts and filling them with our African guys, while also running the training camps in Africa to get more recruits ready to fly over. The business model was strong and I was bouncing back and forth between Africa and Iraq the whole time. Frank and I had another chat about the future, when he asked me whether I was ready to take over. I was getting a sense from him that he was thinking about retiring. I knew he had a good few quid tucked away in the bank and now I knew the business inside out. The time was getting close for the handover. We were on a flight into Baghdad on a business trip one evening when he finally asked me: 'Are you ready to take over the business?'

I told him I was. 'I've got so many ideas,' I said.

'Oh yeah? Like what? What would you do?' he asked.

I rattled off a list of ideas I'd had. I wanted to amalgamate a couple of departments that I thought were a bit flabby. I thought we should relocate some offices to bring the whole company into one secure site to reduce admin and security costs. There were also a few people I wasn't particularly keen on in senior positions, who I thought I'd rather replace with my own people. I was keen to move the business forward, make positive changes. I really respected what Frank had done, but if it was going to be my company now, then it needed to run the way I wanted it to. Frank's mood changed immediately. His face dropped, as though he didn't seem to like most of my ideas; perhaps he felt attached to some of the people I wanted to replace, maybe he felt insecure or criticised by some of the changes that I was proposing. I didn't mean it like that; I was only letting him know, out of respect, the way I was thinking.

Before we could even really clear the air, I was in the bar in Iraq that evening and the news about the Haiti earthquake came on the TV. The pictures of the devastation were truly shocking, a whole country literally shaken to its core. I couldn't believe what I was seeing; the reports said 75,000 people were missing, presumed dead. Even after all the things I'd already seen in my life, and even though I was in Baghdad where people were being killed around me every day, that figure was mind-boggling. A couple of hours later, the news story cycled round again, only this time the death toll had doubled. There were now 150,000 people missing. By the time the full extent was known, the final official death toll would be put at 230,000.

I slept badly that night, as I often do, a result of years of getting up early in the army, operating on little sleep. I got up around 4 a.m. and went down to our canteen for a cup of tea. It was empty except for one person: Frank. I didn't even need to say anything before he looked up at me.

'You're thinking the same thing as me?'

He was right, of course. I was. We were both thinking that we had to do something to help. We didn't want to send money because we knew that money often goes missing in these situations, but we knew we wanted to do something. What we needed was more information about what was happening on the ground. First-hand intel was hard to come by because everything out there had been destroyed.

The SAS is a great club to have been a part of for many reasons, and one of the best of those is that there's always someone you can call on – someone from the Regiment will always know something or someone, anywhere in the world.

Even Haiti. One of the lads from the Regiment, Rog, put me in touch with a doctor I knew had been attached to SAS operations a few times and was apparently now out there doing relief work. He'd been sent out to Delmas in Port-au-Prince almost immediately after the earthquake happened and Rog passed me his number so I could call him directly. I asked him, 'What's the situation out there?'

'The situation here is dire,' he said. 'I've just this minute stepped out from operating on a 10-year-old kid who's lost both of his legs, but we're so poorly supplied that I'm pretty sure he's going to die in the next hour.'

'Well, what can we do? What do you need?' I said.

'We need everything, Billy,' he said. He sounded desperate. 'People, medicines, you name it. Everything. But the thing we need most is buildings.'

I knew that buildings were something we could help with after the experience that we'd built up in Iraq, throwing together camps for the military out there from prefabs. We'd put up some pretty impressive constructions in no time at all. I spoke to Frank and we decided that we could ship forty-seven containers of prefabs from our warehouse in Shanghai, China, to Haiti in six weeks. I'd take a couple of my engineers, use local labour, and put up a hospital and a school building in a matter of weeks.

We knew that if we were going to build a hospital, we had to get the designers on board, so we also got a couple of our guys to start on the designs while I flew into London to get one of the NGOs on the ground to organise being the 'official receiver' of our donation. With their support, it would allow us to clear customs and get the stuff into the country. The deal was that they'd cover that side out of their budget, while we'd provide all the prefabs and find the labour to start the construction.

In the meantime, I decided that I would fly to Haiti to start making preparations and laying the groundwork. The situation there was so bad that only a very limited number of planes were allowed to land, so I flew instead into neighbouring Dominican Republic and drove the eight hours over the mountains. At first, when I landed, I couldn't see what the problem was. The Dominican side of the island was untouched by disaster, still a beautiful Caribbean paradise, with palm trees fluttering along white-sand beaches under a crystal-blue sky. But when I crossed the mountains, day turned into night.

I can only compare the devastation to the battlefield. As soon as we crossed the mountains, the roads on the Haitian side of the island were all smashed, full of potholes the size of craters, as if bombs had been dropped all along them. The shanty towns either side were totally destroyed, the walls either falling down or completely demolished. A three-storey hotel had been flattened as though someone had just smashed it with a giant hammer from above, and there was still a car sticking out of the side of it. People were digging everywhere with their bare hands, looking for the bodies of their loved ones.

There were dead bodies everywhere. They were lying by the side of the road. Bodies that had been dragged out of the fallen buildings were laid out, bloated and stinking; parts of bodies were visible, sticking out of the rubble or burned cars, or squashed along the roadside. Dead animals were everywhere. It reminded me of previous warzones I had operated in. The people who had survived were wandering around like zombies, many in rags, most just totally naked, stunned, shocked, with no water or food or shelter. The stench of death laced the air, made even worse by the heat and humidity.

Once I reached Port-au-Prince, there was nowhere to stay. The few hotels that hadn't been flattened were already full, so I went down to the UN base. I was still officially MoD-certified because of the contract work I'd been doing for the British and American military, and I also had my military ID card because I was really still part of the Regiment, both of which bought me access. Finally, I was given a patch of land where I could pitch a tent. The reality on the ground in Haiti was even worse than I'd seen on the TV. There was so much work to do and I quickly found out how difficult it was going to be to do it. The company in the UK who had agreed to fund the construction of the prefabs that we were shipping to Haiti reneged on the deal as soon as I landed, and so I found myself with enough building materials to construct a small city about to arrive from China, but with no funds to hire the labour to do the building.

It had always been part of our ethos that any time we built stuff, whether it be in Iraq or Libya or anywhere, we employ local people to do the work. There's a lot that the international community can do to help after a natural disaster, but part of that needs to be to help people to help themselves. Jobs are what people need more than anything when their lives have been destroyed, because work brings in money, but also helps to restore pride and self-respect. That's something I'd learned during my time in the army and it was something that I have always been keen to pass on. We were happy to provide the materials, but I wanted local people to do the construction, not foreigners. While I looked around for a new source of finance, I was sleeping in a tent pitched in the UN compound. It wasn't comfortable, but it was in the heart of where everything was happening and so I started to get a better feel for what was taking place on the ground. I began going to meetings, looking for somebody to take our donation, which was already on its way.

A week, then two weeks, then three weeks went by, and I started to feel like it wasn't going to go anywhere. Nobody would take the donation that I was trying to make and the government were now putting a huge import tax on it too, even though we were gifting it to them to try to help their own people. Frank was running out of patience back in Iraq and started to send me messages telling me to write it off and get out of there. It was hard to argue with him logically, and yet something in my gut said, 'I'm not giving up.' So I carried on looking around for ways to make it work. One thing that started to occur to me was that the lack of hotels could be an opportunity for us commercially. I rang Frank to tell him to put another forty prefabs on the sea.

'More?' he said. 'Have you lost your mind? We can't get the first lot in, what are you going to do with more?'

'I'm going to build a hotel,' I said. I reckoned we could build a commercially viable hotel that would generate income for us while I found a recipient for our donation.

I was lying in my tent one afternoon, wondering what the fuck I was going to do to get out of this situation, when I heard a voice from the other side of the canvas that I recognised. It couldn't be. I listened again. It was. It definitely was. I scrambled out of my tent to take a look.

'Billy Billingham?' He shot me one of those winning Hollywood smiles of his and came straight over to hug me. 'What the fuck are you doing here?' It was Sean Penn.

Sean was always one of those crazy cats who'd be getting himself into trouble punching a paparazzo who'd pissed him off one minute, then the next he'd be flying into a disaster area, rolling up his sleeves, clearing rubble with his bare hands and seeing how he could help. He'd done some crazy shit over the years, but right then, in the sticky Haitian humidity, it was great to see a friend. I explained to Sean what a mess our plan to help with the reconstruction of the island had turned into. I told him all about the prefabs that we'd sent over and the big plans we had to build the medical centre that the community needed so badly. And I also took him blow by blow through the circumstances that led to our financing being pulled at the eleventh hour.

'How can I help?' he asked. And then, when I told him I needed a credible charity to take our donation, he said,

'Okay. You got it. J/P HRO [Haitian Relief Organization] will take it.'

Sean already had a hospital project running somewhere else on the island, so we agreed we'd use my prefabs to build a clinic for women who had been the victim of rape and sexual abuse in the community and a school called the School of Hope, on the site of an old golf course that was home to 50,000 displaced people living under tarpaulin. At the same time, I began construction on a 200-room hotel complete with swimming pool, terrace bar and the best restaurant on the island. I wanted something that could be a refuge for the people who were coming to Haiti to work and also to justify to Frank why Sabre should stay out there. I saw a real opportunity to use my business and our project to kick-start the economy and, at the same time, create a sanctuary for the people who were working hard to rebuild the country.

With both constructions running, we were employing nearly 200 local labourers. We trained and developed electricians, plasterers, decorators, plumbers, giving all of them a living wage and a brand-new set of bankable skills.

Sleeping in a tent opposite me on the UN base was an American woman, Jules, who caught my eye. The problem was that I'd decided she was way out of my league, so it took me a while to even talk to her. Jules had started her own global charity, REBUILD globally, which was providing education and job training to some of the most vulnerable in Haiti, which meant that we already had a lot in common, and also that we started to bump into each other more at the same meetings, going to the same events, until eventually I couldn't *not* talk to her. When I did, it was obvious that we were not only going to fall in love, but that we were going to do a lot of good work together.

I could see how passionate she was about helping people; it was brilliant to watch her and her enthusiasm was instantly infectious. She had lots of ideas about how she could start and sustain a business on the island that would give people jobs. We were on exactly the same page in that ambition, so I felt like I wanted to help. As soon as construction on the hotel was finished, she became our first client so to speak, and I was able to keep her close and safe. We found ways to use her philanthropy mindset and my business skills to employ the Regiment's mentality of hearts and minds. Jules was also teaching me about the problems with orphanages in Haiti, specifically that sometimes the orphans were not just children who had lost their parents. There existed a callous sort of business model whereby extremely poor parents were approached by corrupt orphanage directors for their children so they could use these kids to get money from the religious missions that would come to play with the 'orphans'. We didn't want to be part of the problem.

Our relationship grew as we spent more and more time together in Haiti, learning about the conditions on the island and the complexity behind the disaster. We wanted to come up with long-term solutions, but also to use the hotel to bring short-term relief where we could. I was able to acquire items like bunk beds and provide building support, while

Jules identified where we could give them away responsibly. We had the idea to start a Sunday Funday, bringing the kids to the hotel every week to let them play in the pool and just be kids, while the chef made them all sandwiches. Jules had a bottomless appetite for human kindness – I call her the 'Heart on Legs'.

We were an odd couple as far as everyone else was concerned because we were so different. She's American, I'm British. She's from a charity background, whereas I was the military man, but somehow it worked. The people around the place used to call us the 'Tree Hugger and the SAS Killer'. But we didn't care; we'd laugh it off. 'Yeah, I kill 'em and she buries 'em,' I used to joke. As time went on, and all joking aside, we were able to effect real change. I became the ambassador for REBUILD globally and still support the efforts of education and job training, while Jules opened the first-ever woman-owned factory in Haiti, deux mains, which makes shoes from recycled tyres and locally sourced leathers. Nearly a decade later we have more than fifty women employed in the business helping them to help themselves climb out of poverty with dignified employment. Nine of the children we helped get back into school in 2010 are now attending university. One of them is even in medical school – top of his class. An amazing success rate given that only 1 per cent of kids in Haiti ever attend university. Our projects work.

It was a crazy time. Together, Jules, Sean and I were making shit happen. We were creating jobs, putting up new buildings, helping a community to help itself back up onto its feet. Shortly after the hotel was finished, the humanitarian medical organisation Médecins Sans Frontières, impressed by what we'd done, came to ask us to build another hospital, with a specialist burns unit, the first on the island and a multi-million-dollar contract for us. I had to call Frank and tell him we needed 400 new containers of prefabs.

Every day we were overseeing and looking out for new projects, while the nights were always spent hanging out in the hotel bar discussing new ideas. One night, Sean came into the bar with a long face on him, looking really vexed.

'What's up with you?' I asked him.

'I've got a problem,' he said. 'A friend of mine has got into a situation out in Egypt and I'm really worried about him.'

I bought him a beer and got him to explain to me what was going on. The friend was the CNN journalist Anderson Cooper, and the situation was that Anderson was under siege in a hotel basement in Cairo after a protest he was covering had blown up and a bunch of the protesters had started attacking the journalists who were trying to cover it. Sean was old mates with Anderson and had received a call asking if he knew anyone who could help. The story was all over the news; CNN was playing it as the lead story. Sean was genuinely worried about the coverage he was seeing.

'Hang on,' I said, 'I'll make a call.'

I called Tony and he put me in touch with another ex-military guy based out in Dubai, who ran a security company in the Middle East and had boots on the ground in Egypt. We spoke on the phone and came up with a plan and a codeword,

'Gold-Gold-Gold'. Then I got through to Anderson on his mobile and told him what was going to happen.

'Stay locked in the basement for now, mate,' I said, 'but as soon as you hear "Gold-Gold-Gold", then open the door and go with whoever is there. You'll be all right.'

Then I went back to the bar and had a few more beers. To be honest, a couple of hours later I'd completely forgotten all about it, until Sean came back in and started pretending to bow down in front of me, with all that 'we're not worthy' jokey crap.

'What the fuck are you doing?' I asked him.

'Anderson Cooper is on a plane back to the US,' he said. 'Your guys just saved his life.'

'Well, in that case,' I said, 'the next round's on you.'

There was so much to do in Haiti that projects came up all the time. On one particular occasion, Sean, Jules and I decided to support a project by the beach that was set up to house and assist the deaf community. We decided to make a day of it and, after the site visit, we went to have lunch at Wahoo Bay Beach Club. The drive to the beach takes Route Nationale 1, north towards Saint-Marc, through some of the rougher areas of the island, where there's a lot of drugs and gang-related violence. After the earthquake, things got a lot worse up there. As you would expect, disaster increases poverty and poverty brings out the worst in people.

After a few hours, we were back in the car and headed towards Port-au-Prince, driving along the same road. Everyone was in a good mood but something went off inside me as we turned the corner. A lorry had almost jackknifed in the middle of the road, blocking one side, and a commotion was in full kick-off. Lots of people were out in the street, screaming and shouting, and then, from the crowd, three young guys in their mid-twenties started to walk directly towards our vehicle. That's when I saw that one of them was carrying a pistol by his side. I suddenly thought to myself that this was going down. I had to act.

I shoved the car into gear and accelerated straight at the three guys, taking them by surprise and forcing them to jump out of the way. Then I noticed that the lorry had actually started to reverse further into the road, making the gap between it and the wall at the side smaller and smaller. By the time we reached it, we were doing at least 60mph and rode the car up the kerb, clearing the gap between the lorry and the wall by a matter of inches.

'What just happened?' Jules asked me when she'd finally got her breath back.

'Something bad was about to happen, right?' said Sean.

'Correct.'

We got back to the hotel and, while Jules ordered some drinks, Sean took me to one side. He had seen the pistol in the guy's hand, too. We couldn't be sure, but we both had a sense that we had been targeted. We sat down and everyone began to calm down a little until we heard on the news that a German NGO worker had been abducted and her driver shot dead on the same stretch of road, two hours after we'd passed through it. She was finally returned three days later after a ransom had been paid, but not before she had been brutally raped multiple times.

—

The hotel was a great investment for Sabre. It was making money, and we were also catering to the relief workers and friends who were making a positive impact. We had created a shitload of local jobs off the back of it and there was a tremendous amount of goodwill for us on the island. Plus, it was the place to hang out in the evening, which had led to some other good meetings that, in turn, had led to new contracts for us. However, despite the extra work coming out of Haiti, my relationship with Frank seemed to be taking a turn for the worse.

Frank flew out to see how things were running and couldn't believe how well we'd done. Instead of embracing what I had built and all our achievements, his feelings of possible jealousy and negativity were overwhelming. My daughters were there at the same time; they'd known Frank and his family for a long time – as a matter of fact, we were neighbours in Hereford. I was a bit surprised when Zoe asked me one evening what was up with Frank. She could tell that we weren't acting like the best mates that we'd always been. During this visit, Frank and I went to lunch with an influential guy and very successful local businessman, who was building a $30 million hotel on the island. The businessman had long connections to Haiti and he was pretty open about the fact that he'd probably lose money on the project, but he wanted to put something into the place and leave it as a family legacy. The businessman launched into a glowing monologue about me to Frank, all about what an inspiration I'd been and how what I'd done on the island was a great example to other people. I love that businessman, he's a great bloke. But suddenly, out of nowhere, he and Frank had a massive row that got way out of hand. It ended up with Frank and I having to leave. Frank left Haiti on the next flight.

Maybe I should have been worried and aware of what was really going on behind my back. Trust seemed to have been replaced with greed. I could have insisted on our partnership with Sabre being officially written down as a formal business contract. I could have done things differently, but I didn't because I worked on trust between friends. A month later, however, the payments to the Haitian hotel's suppliers stopped.

After that, payments to staff and utilities stopped, including me. We were being slowly cut off. Then a lawyer came and served me with an eviction order. Two months later I was evicted from my own hotel. Soon after that, it was sold and Frank stopped returning my calls. I left Sabre with nothing. Sabre International no longer exists. I've read that various people sued Frank and that some of those claims were settled out of court, and I've also read that a much bigger investigation into Sabre's business activities by agents in the US is pending. I'm grateful for everything I have now and I don't regret anything, and all things considered, I think maybe I did well to get out of Sabre when I did.

CHAPTER 9

Making Movies

My first foray in front of the camera wasn't brilliant. In fact, people could have died. The first show I ever got involved with was a reality show called *Unbreakable* for Channel 5. The premise was that eight super-fit people would follow adventurer Benedict Allen – lovely bloke, ate his dog once – to the world's toughest physical environments to be tested to their limits. These superhuman, elite athletes were supposed to be so fit, so strong mentally and physically, that nothing could break them.

The producers decided to set the first episode in the jungles of Guyana, a nasty secondary jungle full of malaria and dangerous animals in South America, not the kind of place you'd ever want to find yourself alone on a dark night. They called the army to ask whether they could use any of their instructors on the show. Of course, nobody who's still in the special forces can appear on a TV show, but the guy they spoke to, the OC out in Belize and an old recruit of mine, said to them, 'We can't help, but if you want someone who really knows the jungle and is absolutely fucking horrible, then I have just the guy for you.' Guess who that was?

A few days later, my phone rang and the producer of the show explained to me what they wanted. In the episode, Benedict would take the group – made up of a boxer, a martial arts expert, an iron man etc. – and introduce them to 'cultural challenges' like rubbing nasty bugs all over their skin and whipping themselves with bamboo, all of which sounded like a nasty infection waiting to happen, as far as I was concerned. My job was to beast them physically in between, the way they might get beasted during selection. In the end, they wanted to prove that these characters were, as the show said, 'unbreakable'.

'I'll break them,' I said.

'No, you don't understand,' he said. 'These people are elite athletes, they're unbreakable.'

'Okay,' I said, 'but I will break them.'

'No, you don't understand,' he said.

It seemed to me as though he was asking me to try to break them while also telling me that I wouldn't be able to do it. Eventually I just said yes. Fast-forward a couple of months (as they say in the television industry) and I was stood in the Amazonian jungle, waiting for the super-elite athletes to arrive to film our first segment – a 2km hike through the jungle to the place where it had been decided we'd make camp. There were cameramen positioned at various points along the route to get shots of us, after which the plan was to all RV at the camp and set out the activities I had planned for the next four days.

The problem was that it was already getting late and there was still no sign of them. I tried to explain to the producer that we were running out of time and that, to make it to the camp, we should have left hours ago.

'No, you don't understand,' he said. 'The camp is only 2km away. We've got loads of time.'

The thing is, I did understand. I understood because I've spent years of my professional life working on and overseeing ops, as well as running training camps, in jungles all over the world. And if there's one thing I know well, it's that 2km in the jungle feels like 20km outside of it, and that if you haven't reached camp before it gets dark then you're in serious fucking trouble. Benedict and the 'Elite Eight' finally turned up a little after 4 p.m., which meant we had less than two hours before we'd be in pitch darkness being eaten alive.

I would never ordinarily run in the jungle. Running is a crazy thing to do in those conditions, but, with daylight running out, we had no choice, so I set off, screaming at the group to keep up with my pace. If you think the jungle is unforgiving in the daytime, once it gets dark, there are a million more bad things that will make your life very uncomfortable. You don't move in the jungle at night; if you do happen to get stuck out there in the dark, then you just put your hammock up and wait until morning. The problem was that none of those guys had ever put a hammock up, so that wasn't an option. I was seriously worried that someone could die if I didn't get them to the camp.

Heather, the martial artist, was the first to fall. She lay on the dirt, holding her ankle for a while, every minute eating into our time. 'Pain is glory,' I told her. I had to say something. It seemed to work, because she got up again and off we went. I was becoming increasingly worried, because it was getting dark fast and there were lots of places where we had to stop to cut our way through the undergrowth. Any time we got a clear run, we had to up the pace and move it. I was fuming on the inside because this was not what I had agreed with the production team. TV might be entertainment, but the jungle isn't. The jungle is serious, and it will fuck you up if you give it a chance. I'd been put in an impossible position – I had to get those guys to the safety of the camp, even if that meant pushing them to their limits to get them there. And this was day one!

We finally made it to where the camp was set up and I instructed everyone to go and take some water on board. We'd made it by the skin of our teeth. One of the team came running over to me.

'Billy,' she was screaming, 'it's Nathan. He's collapsed!'

Nathan was a professional powerlifter from Manchester. In the opening credits of the show, he confidently states that his mind is going to be the strongest thing that will get him through. Unfortunately, his body had other ideas, because when I found him, he was lying on the jungle floor convulsing. Eventually the medic and I got him to stop shaking but, because we were deep in the rainforest, we had no oxygen. Nathan was running a fever and was quite delusional. All we could do was get some salts and fluids into him and load him onto a boat downriver to the nearest medical facility. That was the end of his *Unbreakable* experience. It took the jungle exactly one day to break him.

Over the next four days, I tried to find challenges that wouldn't create any more Nathans. I pushed the remaining seven as hard as I dared so as to create a bit of drama for the show but not so hard as to put anyone in hospital. To be honest, it was a nightmare, but we got what they needed. At the end of the shoot, the producer came to me to ask if I'd be up for carrying on and being part of the next episode. He wanted me to do my stuff in the mountains and desert etc., but I was pissed off with the whole thing. I wasn't interested in doing something where the people in charge didn't listen to the experts around them and put people's lives at risk. They simply assumed that because these guys were so fit, everything would be fine, but there's a world of difference between the air-conditioned gym and the 90 per cent humidity of the rainforest.

'Listen,' I said to him, 'there are guys who are in the actual SAS who have died in the jungle.'

'No, you don't understand . . .' he started to say.

'No, *you* don't understand,' I said to him. 'I ain't doing this any more.'

After that, I didn't want to do any more work directly in the spotlight, so I said no to any and all offers that came my way. I preferred to stay behind the camera, doing a bit of security or advisory work for shoots around the world, bits of bodyguarding for Tony and the odd security job on a freelance basis. One minute I could be in Libya running a security team, the next I was in Nigeria bodyguarding a politician or in Tanzania catching elephant poachers or Kuwait training the emir's new security team. I always seemed to be flying into or out of somewhere. Every month was a different task from the last. I still had that Regiment mentality, still chasing the buzz and thriving off the challenge of taking on something new.

I was in Kuwait when I saw that my sister was calling me in the middle of the night. I looked at the phone and suspected right away why. I knew it didn't mean anything good. I picked up; she was hysterical, crying, so I said to her, 'It's Mum, isn't it?' Mum had had a cancer scare a few years before that and, even though she'd fought it off and was in remission, I knew it was only a matter of time before it raised its ugly head again.

'No, Mark,' she said, 'it's Dad. Dad's dead.'

If ever I was going to get PTSD, it was in that moment. It didn't make any sense to me. I could not get my head around it. I hung up the call and rang my brother to ask him what was going on. My sister must have made some kind of mistake, surely, but my brother said the same thing: my dad was dead. I flew back to the UK for his funeral. I wanted to do a eulogy for him. I wanted to say all the right things about him to do him credit, but I realised that, when I came to write the words, I had to ask myself hard questions about who he was, who he *really* was. I always described my dad as a ring of steel around the family, but sometimes when I was kid I'd felt too that he didn't have enough time for me, because he was always working. But then I realised that he did that for us, literally always working, every hour, so he could provide for us.

It had taken me my whole life to really understand my father. It's a sadness that it was only after he died that I understood what a great man he was and how he had been such a massive influence on our family. A quiet man for sure, but in many ways the epitome of silent love. For many years I had wondered if my dad really thought I would fail in the army that time he took me to one side before I went down to Aldershot, or whether he was trying to put a rocket up me, as if he was saying, 'Go and prove me wrong.' It went through my head every single day, but I only found out the startling truth much later.

I was packing up his belongings in the days after his funeral, when I found a shoebox in his room. It was full of clippings cut out of newspapers from when I was a kid right through to that day. There were cuttings from the local paper of boxing fights that I'd won, stories about my time in 3 PARA and in the Regiment and accounts of battles that I'd been involved in. I'd had no idea that he'd even seen them. I'd always assumed that my dad didn't care, that he was too busy working to notice what I was doing, but I was wrong; he was watching me the whole time. He never once talked about it, but now, only after he'd passed, I knew that he knew. I was numb with shock and also at the same time proud as any son can be of his dad. After the funeral, Jules and I flew back to Florida, where she was living, to take a little time.

Three weeks later, the phone rang again. My mum had died. Again, I was absolutely devastated. After everything that had just happened, I couldn't believe I was now facing exactly the same trauma and another huge loss. I thought about what my mum and dad would want for all of us and I knew that, ultimately, they only wanted us to be happy. They would have wanted us to pull together and stand strong. It was a devastating loss but, as the weeks of trauma and sadness went by, I focused more on how lucky I was to have had them in my life as role models, and that's what got me through. Not a day goes by when I don't think about them both.

—

I went back to working freelance jobs, keeping myself busy, taking on whatever came my way, flying around the world, keeping my head down. In between trips, I was flying back and forth to see Jules in Florida or Haiti, where we still had the factory and a shop as part of a turnkey project. We were still running the charity out there despite the debacle with Frank and the hotel. I was in Hereford, packing for a trip to go over and see her, when I saw a familiar name come up on my phone. It was Sean Penn.

'Billy, I need a favour,' he said. 'I need you to be in Barcelona next week. I'll explain everything to you then.' I'd do anything for Sean. Love the guy. But what he was asking was too much. I hadn't seen Jules in over a month and I'd already arranged to fly over to meet her in Florida, so we could spend some time together. No matter how much I loved Sean, I loved Jules more. I told him in no uncertain terms that I couldn't meet him in Barcelona.

'Jules likes Barcelona, doesn't she?' he said. Typical. Won't ever take no for an answer.

'I don't think she's been,' I said.

'She's gonna love it,' he said. Then he offered to fly her there and put us both up in a nice hotel, all expenses paid. I said I'd run it past her but not to expect anything. Turned out he was right, because the following week I was on my way to Barcelona with Jules.

On the way in from the airport, I got an email from Sean. It said: 'Read this and tell me what you think.' I opened the attachment and couldn't believe what I was looking at. I thought there must be some kind of mistake, that he must have lost his mind, because attached to the email was a script, the script for Sean's new movie, *The Gunman*, and in the script was a part . . . for me.

I can't do this, was my first thought. And then, *Why can't I?* was my second.

The movie was about an ex-special forces guy running a team of contractors out in the DR Congo, who gets involved in a corporate-sponsored hit on a mining minister, only to find years later that his past has caught up with him and his life is in danger. The cast was a who's who of some of the best actors in the world. As well as Sean, there was Javier Bardem, Mark Rylance, Jasmine Trinca, Idris Elba, Ray Winstone and some bloke called Billy Billingham (Sean called me Sir

Billy Billingham in the credits). Principal photography was in Barcelona. It meant that this trip to Barcelona wouldn't be a flying visit. We'd be living there for nearly three months.

As well as playing the part of 'Reed' in the movie, Sean wanted my advice on practical stuff like gun drills and special forces terminology. How would they act in this situation? What language would they use to describe this operation? What kit would they use to do this task?

When I got onto set for the first time, I realised that I was the eighth actor on the call sheet. I wasn't some bit-part or an extra; I was in loads of scenes. I had scenes with Javier Bardem and Mark Rylance, for fuck's sake. It was nuts, it was just surreal. That's how it felt all the time, every single day. I kept wondering when Sean was going to crack and admit that the whole thing was just a piss-take. I even had my own trailer. A runner came down every morning bringing me coffee, bringing me teas. Jules kept being given flowers. I got picked up at the hotel every morning by my own driver and our room in the hotel was the size of my house. It was just ridiculous.

Every day on set, I could see all the bodyguards; some of them were lads I knew from when I was doing security for actors. They were looking at me, asking themselves, 'What's he doing here? Is he an extra?' Everything about it was comical. Before we got going, we had a press day in front of the world's media to start a buzz about the film. I didn't have a clue what I was doing or what was going on. I was sat up there in front of the cameras, answering questions about the movie with all the other actors who were in it. I was talking about my character, while Idris fucking Elba was talking about his, and he was looking at me, not knowing anything about me, probably assuming I was another actor, like him. The one person, other than Sean, who I knew was Ray Winstone. Ray and Angelina Jolie had worked together years before on *Beowulf*, while I'd been doing her security, and Ray had often come into the bar for a drink. As the token Brits, we'd hit it off, and so, any time our paths crossed after that, Ray would always come up and say hello. But when he saw me acting in the same movie as him, he was like, 'What the fuck?' It was hilarious.

It was hard to get my head around what I was supposed to be doing. My character, Reed, a former special forces guy, was now working alongside Sean's character. That much was fine, but the technical side of the process was all new to me. One day, I was lying back in my trailer, having a nice time, while outside the whole crew were getting ready to start rolling. Every second on a movie set costs a fortune, so you've got to be pretty sharp when they're ready for you, but a lot of the time is spent waiting for that call. Over the radio in my room, I could hear them calling 'Reed' to set, over and over. I just lay there, chilling out, relaxing, happy that it wasn't my time to shoot yet.

The next thing I knew, the runner came knocking, sweating his ass off, flapping at my trailer door. I got up and let him in. 'Reed?' he said, in a blind fucking panic. I didn't know what he was talking about. I was about to say, 'No, I'm Billy, mate,' when I remembered that was the name of my character. 'Reed? Oh fuck, yeah, that's me.'

By the time I got to set, Pierre Morel, the director, was fuming. All the other actors, crew, everyone, were stood there, waiting for me, not able to do anything the whole time I'd been sitting on my arse in my trailer. As I came running onto set, Sean piped up, 'That's another fifteen grand.' I felt sorry for Pierre trying to get me to act sometimes. There was a line I had that was in French, which I couldn't remember for the life of me. He literally took me aside and, for about forty minutes, one on one, went through it over and over until finally I had it. Poor bloke. We got back into it and, as soon as my cue came up . . . blank, nothing; I just could not remember it. Pierre came running back, screaming, pulling out his hair. 'Fuck it, fuck it, just scrub it.' The line was dropped.

The truth is that I was a bit overwhelmed by the film set. I didn't know what it was exactly, I'd been on film sets before, but I'd never really looked at them that way, from the inside looking out. I found the constant stop-start repetition really hard to deal with. Every shot was done over and over, every line repeated and shot from four different angles. One scene that Mark, Sean and I did in the back of a van took two days to shoot, even though it was just two minutes of the film. In the scene, I had to pull down my trousers and show the boys my arse. I was going crazy by the end, so I told Jules during a break that, on the next take, I was going to trick the boys by getting some toilet paper and hanging it out of my arse crack like I'd accidentally left it there after having a shit. I didn't do it, of course, I was just joking, but Jules, watching on the monitor on set, was so convinced that I was going to do it that, a second before I pulled my pants down, she screamed out, 'Noooooo!'

The next thing I heard was Pierre shouting, 'Cut!' And then Sean: 'That's another fifteen grand.'

—

The following February, on a cold, wet London evening, the whole cast was called to see the film for the first time at its world premiere. It was a full-on red-carpet gala event at the BFI cinema on London's South Bank. I went with Jules, in my sharpest suit, making the most of the fact that this was probably going to be the first and last time I'd ever walk up a red carpet in front of the world's press to see myself at the premiere of a major Hollywood movie – the last time I'd face a wall of paparazzi cameras flashing at me alongside the likes of Idris and Sean and all the other superstars in the cast. I'd already said to Jules that I didn't think this was the thing for me. I certainly didn't have any ambitions to be the next Vinnie Jones or anything like that. The movie-making process simply wasn't my thing. It was just a bit of fun, a one-off; I needed to be more active. 'Hurry up and wait' wasn't my thing.

It was hugely nerve-racking as we took our seats inside a packed cinema. The size of the operation suddenly hit me as I looked around at all the people who'd come to see the movie. The second I saw my own ugly mug up there, 20ft high on the screen, was full-on. I cringed, couldn't help but critique myself the whole time, thinking how I could have done that better or said that a different way. I hate the sound of my own voice.

But gradually I settled into it, recognising the craft and skill that had gone into making it. Sean had ended up taking over a lot of the post-production and I was really impressed with what he'd done. The film was actually pretty good. Suddenly I didn't mind so much the six hours shooting my arse in the van for the two seconds of the movie that it made up. It all seemed worth it now. It actually felt good. I started to enjoy it, getting into the plot and appreciating everyone else's performances. In the bar at the afterparty, everyone was pumped, telling me how I was a natural and how I should do more acting. I loved it, but I also knew one thing – *I ain't no natural.* I enjoyed seeing the final results, but there was no way I was going to become one of those guys who reads scripts and goes to auditions and all that shit. Just wasn't me.

Sean came over to me, really ripped and full of the buzz. He took me to one side for a quiet word.

'Listen, Billy,' he said, 'I know how the sequel starts. It opens with my character walking through New York during the St Patrick's Day parade.'

'Right . . .' So far so good. I wasn't sure why he was telling me.

'Then he sees Reed in the crowd!' Hang on a second. That was me. I knew my character's name now. Sean Penn was offering me a part in the sequel to his movie. You know what I said about not wanting to be a movie star? Ignore that. I'm up for it. Of course I'll be in your movie, Sean Penn. When do we start?

(I'm still waiting, Sean.)

CHAPTER 10

Up to the Tusk

When I first got the call in 2009 from my mate Andy saying he needed someone to fly out to Kenya to help him set up an anti-poaching team there, I jumped at the chance. I've always loved Africa, absolutely adore it, can't get enough of the place. If you've ever been then you're probably the same and a part of your heart will for ever be there. It is such a huge continent that you cannot appreciate the enormity of it until you're right there, seeing how city and urban life can suddenly make way for wild and dangerous fucking countryside that can spring an attack – from both man and beast. There was always something exciting happening for me when I was there and despite seeing how many of its people were stricken with poverty, for an ex-soldier like myself who was keen to ply the skills he learned in the British Army to make a difference, it offered a huge challenge. Dropping into such an alien environment didn't faze me at all. I loved the names they have for everything; I love the way the people greet you and live their own lives, the passion that they have for what they've got, which is next to nothing most of the time. It's just a great part of the world to be in.

The job I was hired for initially was to work for Kuki Gallmann, an Italian poet, who'd moved to Kenya with her husband in the 1970s and set up a huge conservation project there. Since then, her husband had been killed in a fatal car crash and her son by a deadly snakebite, but Kuki had persevered, building up her 100,000-acre ranch and using it as a base from which to spread her message – that people and nature can operate in harmony in Africa. She had dedicated her life to educating local people, while at the same time protecting the animals and habitat on her land from the dangers that threatened them, the most pressing of which was poaching.

I arrived at Kuki's ranch, Ol ari Nyiro, at the end of the rainy season, when the land is in full bloom and shockingly beautiful – it takes your breath away. The lands that surround her ranch were luscious and green, endless rolling hills and savannah, the perfect habitat for animals, which is why Kuki could boast to have not only a huge population of elephants and lions on her land, but also two of the last black rhinos in existence anywhere in the world.

Our Land Rover pulled up outside the house, a sprawling two-storey, ex-colonial-style African ranch, hardwood pillars supporting an enormous thatched roof that hung majestically over a teak deck from where you could sit and enjoy a cold beer while you watched the elephants enjoying their midday swim in the lake further down the valley. Kuki had built a whole area with a pool and guest houses for her daughter's wedding that hadn't been used since, which was perfect for Andy and me to use as a base while we got to work.

We had already collected a little local intel on what the immediate dangers were to Kuki's animal population. The rhinos were the jewel in the crown and so had to be protected at all costs, but the elephants were being increasingly targeted by local poachers, making money from selling tusks to the Chinese for tens of thousands of dollars a pop. The problem we faced was how to take on what seemed like an army of poachers with only a handful of local guys in an area almost a quarter the size of the whole of Herefordshire.

The men we had were a little rag-tag bunch. In total we had ten fulltime guards who hardly had a pair of boots between them. The first time I met them and lined them up with their equipment, my heart sank at the scale of what we had to achieve. They were all local tribesmen, and, although most spoke some English, none had any military training or background. When I spoke to them about their roles, I got an instant sense of their passion. They were really genuine people who wanted to make a meaningful contribution, but we had our work cut out to get them up to any sort of effective standard of operation.

I had brought out uniforms and boots, generously supplied by the UK military. Andy had managed to source a few additional weapons, a couple of old AK and an old 303, but that was it. We didn't have any other weapons beyond machetes and clubs.

We began taking the men through some very basic training, starting with stressing the importance of dressing smart, looking and behaving like a military outfit. Then we started teaching them how to do tracking, how to follow a trail, how to use the weapons they had. We taught them how to react to a situation, how to do observation, building observation posts in strategic positions around Kuki's land. We taught them how to carry out aggressive action, should poachers come into the area, and we had to instil a confidence in them that they could fight fire with fire.

The whole time we worked, more and more elephants were still getting killed. It was such a vast area that while we worked on training, the poacher knew they could exploit us further and enjoy more freedom of movement. They started really going for it and it seemed like every day a fresh elephant carcass would appear on our morning patrols.

One morning a delegation of British dignitaries was visiting the ranch and Kuki asked us to do a survey of the land. She was keen to put the best face on for her guests, but the problem was that night, three elephants had been taken down. The body of a dead elephant is a tragic sight to have to see. Full grown, they stand around 10ft tall and weigh 5,000-6,000lbs. The poachers only take the tusks, using chainsaws to cut through the ivory as close to the root as possible so that half of the face is often missing. It was horrifying. I just couldn't believe that people could be so cruel or so desperate to do something so barbaric.

The three elephant carcasses we found that morning were full-grown males and had attracted the local lions. By the time we reached them, the lions were already picking the flesh of the bones, in no mood to be interrupted, which meant that we couldn't get close enough to move the bodies away. Kuki's VIPs were just going to have to face the reality of the situation. At this rate, the future of the African elephant was at risk.

Andy and I quickly understood that it wasn't the poachers we should have been going after. They were just poverty-stricken people who'd been offered a bunch of money that they weren't in a position to turn down. We realised that we needed to operate on two levels: we had to hold back the poachers and at the same time begin to sniff out the links to who was funding them. We began to ask questions of the people in the area about anything that seemed suspicious, who looked like they were from out of town, who didn't fit in, etc. We knew that they were bound to be foreigners, so we were trying to make links through local villages to find out if anyone had seen anybody who looked white or Asian, but it was difficult to get information – everyone seemed scared to talk. In the meantime, elephants kept dying, so we had to focus our efforts where we could – on the poachers themselves.

The one thing we did know was what paths the poachers were taking into Kuki's land. We were able to track their routes, so we began to build observation posts in strategic positions. We put on a show of force that suggested we had ten times the men and arms that we really did. We ensured that, any time the ranch got a visitor, every time a group of tourists came onto the reserve, we had all of our best soldiers, guns and ammunition on show as a sign of strength. We built way more observation posts than we could possibly man to make it look as though our numbers were far greater than they actually were. Obviously we still had nothing in terms of real fire power compared with the enemy, but we didn't let anyone else know that. As far as the public perception was concerned, we had built an anti-poaching army and we were ready to take on and take down anyone who came for our animals.

Our one major asset that the poachers didn't have was the British military mentality – the ability to think outside the box, think around the problem and, in particular, create the illusion of strength. That was something invaluable that I'd learned during my time serving with 3 PARA and then in Iraq.

—

The task that elements of the Allied forces were given in Iraq was to find and immobilise any weapons of mass destruction (WMDs) that Saddam was holding there – something that had been well publicised prior to the start of the conflict. The British mission was clear: go in, locate them and destroy them. As has been documented many times in the press subsequently, all the intelligence we received was coming from the highest level, so we had no reason to believe that it wasn't accurate. As far as the British forces were concerned, they had being given a vital job and now was the time to get that job done and prove the assertions that Saddam had these WMDs.

Like all British troops there at the time, it felt great to be finally going to war, all the lads together, pumped and primed, not knowing what to expect, but feeling ready for anything. Intelligence had told our planners to prepare for being gassed and bombed, and British troops were told that it was quite likely that they'd be targeted with chemical weapons too, so vehicles were equipped with warning devices to let them know if a chemical attack was coming their way. On top of that, everyone was kitted out with full respiratory gear, which added to the physical challenges being faced. Our units in the frontline tasked with reconnaissance and detection work knew that the challenge was going to be a force much larger in number than themselves, as the Iraqi army was huge and contained their infamous Republican Guard.

There were no tarmacked roads into the Western Desert, so like many others, our convoy often had to stop and dig the vehicles out of the sand with shovels, clearing paths, laying tracks for them to pass. It was hard-going, non-stop work for over twenty-four hours for British frontline units just to get to the border, where the enemy were waiting for them. For the majority of British troops in all units, this would be their first face-to-face engagement against an enemy since they had joined up. As the British convoys pulled up to take cover, the jets, A10 Tankbusters and Lynx helicopters went up, roaring overhead, dropping fire onto their targets on the other side of the border, trying to create a path for the ground forces. But there were still skirmishes on the ground, open firefighting going on through the night, rockets firing at us from every direction.

It would have been surreal, weird, because they would have rehearsed that moment over and over in their heads a hundred times before, but still nothing prepares you for the reality of somebody running towards you and shooting at you, trying to kill you. I would have been like everybody else in the British Army that was going into this war zone in that I wanted to see how I would react in conflict – to somebody actually trying to kill me. A question we all ask ourselves prior to going into armed conflict for the first time is, *How the fuck am I going to deal with this?* Yes, you might be aware that people see you as a good soldier, or that you would make a good leader or do great things, but right there, in that moment, none of that matters – all that matters is that you live, which means that the person running towards you needs to be taken out – killed or otherwise. That is war. It will be the same for the guys around you, too. Some of the people who you'd think you can really rely on will prove to be not as good as you were expecting, while others can step up to the plate and produce the results. In my wartime experience, some of the quietest men turn out to be the best under the pressure of combat. If you are functioning properly and with a clear head, you don't have to think about what you are actually doing, that you are shooting somebody, because you start putting into motion everything you have been trained to do for many years. It's been said before, but it can really feel like it's either you or them. With myself, it was always a thought process of *This is what's got to be done.*

Once British forward units tasked with locating and destroying Saddam's WMDs broke across into enemy territory, they split up. Half would swing round to the western side of the desert to start manoeuvring forward steadily, watching, looking for locations and targets. The other half would advance towards bigger Iraqi towns 500 miles away from the border in the direction of Baghdad. The problem British forces faced was that there were a hundred tiny villages between the border and Baghdad, and every one of them was a potential ambush waiting for them. Each one had to be cleared and checked before we could advance any further. British forces only had a few specialised units to cover the whole of the Western Desert, 500,000 square kilometres, search all those little places and try to find the launchpads, the WMDs and the chemical weapons caches. Not an easy task.

For the next two days British forces operated a leapfrog strategy. While half pushed forward to a particular village, took it and held it, the other half moved forward to the next village, doing the same. It goes without saying that they would be getting fired at all the time. Every position taken up would have some sort of conflict, some kind of battle, whether it was a stand-off, long-range shooting, taking out positions with heavy weapons, or co-ordinating the jets above. It would be fully armed combat all the time as gradually the British troops leapfrogged their way forward.

Psychological warfare was often the most important tool the British Army had when taking on such huge forces as Saddam had. No one knew at the time that large elements of his armed forces were badly trained and supplied. But if you are a small force wishing to take out a much bigger one, using the terrain to your advantage is the key. This has been the British way since the desert war against Erwin Rommel. The lead units of the Allies would have been small in number, armed to the teeth and supported with massive air power – which would blow the fuck out of the Republican Guard later on. Taking on and defeating a superior force takes ingenuity and balls. You want to deceive the enemy as to how strong you are when there might well be only a handful of you – no more than a platoon. You hit the bastards with everything at your disposal from multiple points at the same time to confuse and put the fear of God into them, until they're actually saying to themselves, *Fucking hell. This is a brigade coming at us.*

The Allies' objective had been to find WMDs and take control of Baghdad, overthrowing Saddam and his henchmen, but now people were talking about pushing out and going further, which to me didn't make any fucking sense unless you were thinking of going into Iran, too. A few days later, British troops received word that Baghdad had fallen, the Iraqis had surrendered and they wouldn't be advancing any further. The whole thing had lasted just six weeks from start to finish. Saddam had been toppled and the people of

Iraq had been liberated, free now to form their own democracy and build a new nation. The plan had always been to pull out and leave them to it. British forces returned to the UK with the expectation never to see Iraq ever again, but many would be dragged back there as the situation spiraled out of control once Saddam had been executed, his regime dismantled and anarchy reined.

—

After our success of employing the 'Iraq strategy' in Kenya, I was asked to return to Africa to establish a much larger anti-poaching force in neighbouring Tanzania. The operation was being overseen by a mate, Tim Spicer, CEO of the AGES group for whom anti-poaching was a real passion project. He wasn't doing it as a profit-making thing but more out of a real love of the animals that needed protecting. He sent me and another British military guy out there to train their anti-poaching team.

Unlike in Kenya, we had been promised a real budget and a sizeable force to train up and protect the animals in Ruaha Park, the largest national park in east Africa. The park was home to good-sized populations of big cats, hippos and buffalo and was once known for having one of the largest elephant populations in the world, before poaching decimated their numbers in the early 2000s. Our mission was put a rapid stop to that decline.

Again, we began using and adapting the skills and training that we'd received in the British Army to bring skill sets to the local gamekeepers that they could begin to employ. We worked on everything from first aid to tracking, showed them how to patrol, to plan and run operations, to engage and defeat enemy combatants. It was like being a DS again, taking raw materials (men) and turning them into quality products (soldiers).

We began to restructure their patrols, getting them to focus on ground sign awareness, tracking, weapon handling, first aid. I treated it as though I was teaching a military-type course. I wanted it to feel full-on to them, exactly like the army training I'd done does, sending them out on team tasks, encouraging them to show initiative and leadership.

It was hugely rewarding for me personally to see how we could take young men who had lived their entire lives in small, rural villages and train them to take the initiative to protect their own land. I always insisted that we recruit from the small farms, the poorest villages, to find people of the right age that we could bring in. My attitude was always, 'Let's train them, let's educate them to, let's show them that this is yours, this is your land, these are your animals.'

Over the months that followed we began to see real change. I could see a great shift in the people being taught for the first time, being given responsibility for the first time as I had. We had a team of the best British military experts giving them training, the best academic guys teaching them about the animals, and slowly we began to see in them how much they wanted to do the job, for the sake of pride, and not just for a job. They wanted it because it was theirs, and they'd never been shown or told that before, never been given that opportunity.

Where we'd been used to finding an elephant carcass a day on our patrols, slowly it became a carcass a week, and then a carcass a month. Because the number of elephants being killed dropped, so too did the number of lions and other big cats, which often get caught in the crossfire of the poachers.

It was a really good feeling to be able to be a part of that, of something that was effecting real change in the park and in Tanzania. It also reminded me of how much I had benefited from my own military training and how great it felt to act like a DS again. I'd almost forgotten how much I loved training people and how much personal pride I could feel from being part of someone else's transformation. I didn't know it then, but it was a sign that my time being a DS wasn't entirely behind me. It would take a TV show to really bring that fact home.

CHAPTER 11

Television's Mark 'Billy' Billingham

Before Channel 4 ever produced the first series of *SAS: Who Dares Wins*, I was made aware of it by my mate Colin Maclachlan, who was the only SAS guy on the first series of the show. I was actually Colin's DS when he came for SAS selection, so I knew a lot about him and respected him, but when he contacted me about the show and asked if I'd be interested in getting involved, I had serious doubts.

'It's a bunch of SF guys,' he said, which sounded interesting, because I assumed that meant they were all going to be SAS, but then he told me what the show was going to be called.

After my experience with *Unbreakable*, and seeing a few of the other *SAS: Who Dares Wins*-type shows that had come out since then, I was bit down on the whole telly thing. In the end, I just felt that I didn't want to be part of anything like that again.

'Thanks for the offer, mate, but I think I'll pass,' was my response. About a week later, another friend of mine, who worked with the BBC, called me up.

'Billy, would you be interested in doing a programme?' These bloody things were like buses. This show was being presented by Freddie Flintoff and was all about the search for the country's 'toughest recruit' in a show for BBC Two. It was called *Special Forces: Ultimate Hell Week.*

The idea was to take the fittest men and women from around the country and put them through a selection-type experience, but the twist was that different SFs from around the world would be represented. Former DS from the SAS, US Navy SEALs and Russian Spetsnaz would all be working together to design the ultimate test of mental and physical strength – basically a selection of gruelling obstacles from all over the world, set over a period of days. Because I knew these lads better and had a good impression of Freddie, I said to them that I'd take a look at the pitch and think about it. In the meantime, I went back to work in Nigeria, bodyguarding for a senior finance guy from Ernst & Young. That's when I thought, *Why not talk to those guys? It'd be more fun than this.* So I called them up and said I'd come in and talk about it. The problem I had was that before I even left Nigeria, I was starting to feel not quite myself. I had a bit of a temperature, aches and pains, headache – all warning signs, having already had malaria three times before, that I might be getting another bout of it.

By the time I got back to the UK, I was feeling really unwell, but the meeting with the *Ultimate Hell Week* team was scheduled for the next day, so I decided to go along anyway. I didn't want to miss being involved in the big sit-down round the table where they talked about how the programme was going to run and what ideas everyone had. After the experience I'd had on *Unbreakable,* I felt strongly that there was no way I was going to do it unless I had an input into how the show was set up. I drove down to Crickhowell, a big British Army training centre, where the filming was going to be based, still not feeling great but looking forward to meeting Freddie and the other guys. We all sat around a big table, sharing ideas and thoughts about the show. The producers and directors were sat along one side, Freddie was there, and we were quickly chatting about ideas, getting on really well, and I liked what I was hearing. I was firing off some ideas for tasks and challenges that I thought we could include, when I realised that everyone in the room was staring at me. Not because of what I was saying, but because of how I looked. My whole head was sweating, thick greasy sweat started pouring down my face; I was really not feeling good, and was suddenly dizzy and weak.

'Are you okay?' One of the producers got up to get me a glass of water. I felt like I was hallucinating.

'Actually,' I said, 'I don't think I am, no.' The next thing I knew, I was in Hereford hospital.

I don't know how I got there because I was so in and out of it, but when I came around the doctors were obviously worried. They thought I had Ebola, so I was put in an isolation ward and they started me on the drugs. Despite all the shit they were pumping into me, I just got gradually worse and worse. As I lay in my bed, I realised that I could smell myself. I smelled like I was dying. I could smell my nan. I remember when my nan was dying of cancer, I used to walk into her house thinking that I could smell her. Literally smell her rotting from the inside. Now I was in my hospital bed thinking, *This is what it smells like, exactly what it smells like. What is wrong with me?* I was in so much pain; it was just horrible, and at times I actually felt like I wanted to die. They did more tests until they eventually decided I had cerebral malaria. The bad malaria. Once they knew that, they were able to start turning it around, but it still took me weeks to get back on my feet, months to get back to full fitness. I missed the BBC show, although I suspected that whatever 'Hell Week' those recruits were being put through, it was nothing compared to what I was going through.

I'm not really a big watcher of television shows, so I didn't see either the BBC's *Hell Week* or the first series of Channel 4's *SAS: Who Dares Wins*, but I heard on the grapevine that the Channel 4 show had gone down well with the viewers. To be honest, I was pleased for Colin and happy for the Regiment, because I liked hearing good things about a show that had SAS in the title. As the series had gone down so well, I wasn't really expecting to hear any more about it, but again I got the call – this time from Andy, one of the producers of the programme. He said that Colin was leaving the show, but they wanted another SAS guy to replace him. In part, it was because they wanted four guys up front, but it was also because the other three guys they already had were all SBS (Special Boat Service). The SBS guys had all passed SAS selection but had never served in the SAS and the producers felt they wanted at least one SAS guy on the show to justify the title.

Right away I said to Andy, 'I'm not interested.' I couldn't understand why they'd want me when there were a load of other people around who could do it. I had a suspicion that it had something to do with the work I'd done with Brad and Angie and how they could generate some publicity out of that, and I didn't want to capitalise on that kind of thing at all. Still, Andy insisted that the reason they wanted me was because of my experience, not only as an SAS soldier but also as a DS. He seemed to have done his research and he knew all about me. In the end, I said, 'Listen, mate, there's no way I'm coming to London.'

'Well, then, I'll tell you what,' he said. 'I'll come to Hereford.' Three hours later he turned up in Hereford at my apartment. We talked some more about the show and he left me a USB stick with the first series on it. I agreed to take a look at it and to have a think about his proposal. I was impressed with the effort he was making.

The next day, I flew back to Nigeria on another body-guarding gig. On the plane, I remembered the USB stick, so I put it in my laptop and, by the time we touched down in Lagos, I'd watched the whole thing. I'll admit, I had some trouble with all the shouting and screaming, because we never shouted like that when I was DSing for real on selection, but I could appreciate that the reality wouldn't work for the telly. You must have noise and colour to hold the viewer's attention. What I really liked about it was that it wasn't all about the presenters. I liked how it was more about the recruits and why they were doing it, not all 'Look at us special forces guys, we're fucking this and that.'

I liked how the DS's role was more about listening to the stories of the people, questioning them, digging deeper into their souls to get to the real core of who they were, their characters, to give them advice based on real-life situations that SF soldiers have been in. I started thinking, *Well, I know how I would do that, what I would do in that situation.* I knew all about how to push people, while remaining firm and fair. Of course, I can also be pretty fucking nasty when I need to be, to get the best out of people, but I think I know where the line is, because selection isn't about breaking people, it's about finding and building people.

Having been a DS on selection, I also found myself guessing who was going to make it to the end of the show and who wasn't. I felt that I could tell from my own experience how each character was going to end up. I'd seen all those guys before on selection. The cocky bigmouth who would be there for all of two minutes; the guy who looked like a male model, too worried about how his hair looked than getting stuck in. I watched the show thinking, *He's going to be gone,* delighted when he was.

I really got into it. I was watching the big guy, John, who eventually won it, and I could see that, although he was confident and fit, he just couldn't keep his mouth shut. I could tell that he wasn't beyond help and, even though I imagine half the people watching the show must have wanted them to throw him off, I thought he could turn it around if he learned to keep it zipped. The DS inside of me was thinking, *That kid's got something. Just like what happened to me, you just have to work on him. You just have to work on what his big problem is. His big problem is his mouth.*

By the time I'd finished all five episodes, I thought that maybe I did have something I could offer to it, that I could bring something extra to what they were doing, something that could help the people on the show. I'd been a DS. I'd done it twice, in fact, first as a young instructor at the depot and then again at a very advanced stage in the SAS. I landed in Nigeria and I'd totally changed my mind. Now I was thinking, *Actually, I might want to get involved in this.* So when I got another call from Andy, I said I'd meet him again to talk about it.

I went down to London when I got back and met with the production team. We spent the time chatting about their ideas, and I was impressed with how involved they wanted all the DS to be in the process. They asked what I thought about everything, how I would do this or that. I went through a lot of ideas with them, from what sort of challenges to do, to what locations to use. We talked about what we could do to test confidence, trust, survival, and I was impressed with how well it had all been researched. It was a million miles from my previous experience in television.

I was also happy to find out that series two would be set in the jungle.

The next thing to do was to meet Ant, Ollie and Foxy. Obviously, this was their show and it was important that we all got on, but I didn't know a lot about them because they were all SBS guys. I'd heard a little bit about Foxy because we'd served in the same places around the same time, but I'd met so many guys over the years and I couldn't remember exactly if our paths had ever crossed.

I was aware of how I was going to come across from their point of view. Because I was older than them and more experienced, I didn't want them to think I was coming in to try to take over, to try to become the chief instructor or throw my weight around. I'd be happy to just slot in and play the part of a DS again. What I wanted was to have a bit of fun and enjoy it, have some good banter with the team and make the most of the chance to feel almost what it would be like to be back DSing in the Regiment again.

Right away the blokes were brilliant, welcoming, helpful, and I just fitted in. I might have been the new boy, the oldest new boy in town, the old man of the new kids on the block, but straight away we hit it off and they brought me in. Once again, I was part of a brand-new team.

—

Series two of *SAS: Who Dares Wins* was set in the Ecuadorian jungle. The production team had taken over a real Ecuadorian military base, which consisted of a number of corrugated-iron-roofed huts set around a square parade ground, enclosed by a perimeter fence – our home for the next two weeks. The way the show works, twenty-five recruits from all walks of life come in to attempt to pass a condensed version of SAS selection over a two-week period. The DS take them through a series of demanding exercises to test their physical and mental strength. Poor performance is usually rewarded with 'beastings', extra physical exercises, which could be anything from fifty press-ups to carrying a 50lb log up a steep hill. In between all of that, the DS routinely call recruits in for cross-examination in the so-called 'mirror room'.

The mirror room quickly became my favourite part of the show. It's essentially a shed on the base that acts as a mock interrogation room. The production team have installed a two-way mirror so that, while two DS conduct an interview inside with the chosen recruit, the others can watch what's happening from the other side. Recruits voluntarily withdraw as they find that they can no longer hack it, and, in the end, we pick the best recruit from those who remain as the winner.

I was looking forward to getting going, because the jungle was still the place I loved most. Having spent so much of my life in the jungle, I've learned to be comfortable with it and I don't mind being wet and sweaty, stinking, dealing with creepy-crawlies all over the place. In fact, it's a second home. I love it; it's fucking brilliant. I felt good the second we landed in Ecuador, glad to be back in the thick of it. I had to quickly get used to the way the show was produced because all the other lads had done it before. In particular, I had to get my head around the 'fixed rig'. The show is filmed in two ways. Everything inside the perimeter is 'on-rig', which means that all the cameras are static, remote-controlled from an office outside the perimeter and placed so they can capture everything 24/7. Because they're all operated remotely, there're no cameramen to be seen. Unless we go out on an exercise, we never see the crew.

The only time that we would meet the producers was when we came back to the camp to discuss the events of the day, or if we got together to let them know what we'd decided to do the following day. There was no script, so we were always free to focus on whoever we thought needed it. We were always able to say, 'That number seven's a bit chopsy, isn't he? Let's work out how to target him the next day.' Just like we would as real DS in the army. I was aware of all the fixed cameras to start with and that I was miked up all the time, but as I got more and more immersed in the action, I gradually started to forget they were there. In fact, I knew I was fully submerged in the experience when I found myself going for a shit and taking the chance to call home. Next thing I knew, I was sitting on the bog having an intimate chat with my wife before I realised every fucker in the production office was listening to us.

In many ways, it was harder for us than it was for the recruits, because we had to be up before them, asleep after them; we did everything that they did outside of the camp, as well as the 'beastings' and the mirror-room interviews, so we didn't get a lot of sleep. People often assume that the DS are all sleeping in a comfy hotel around the corner, but all four of us were sharing one bunk room, sleeping just the same way as you see the recruits on the show, on an old army camp bed with a mosquito net around it. My first thought when I arrived and saw it was, *Fucking hell, I'm back in the army, sleeping in a bunk next to Ant and Foxy and Ollie.* I started to get to know the other lads well. They are all superb blokes, and each of them has a great set of skills, even though we're all very different people. It works for television.

Foxy's a grafter. He's a smart guy who's been through real trauma, so he's a great advert for how you can overcome your demons and still fight on. He's got a massive heart and I think he genuinely enjoys helping people who come on the show, which is probably why everyone likes him. Ollie is probably the wisest of us, the thinker who gets deep into things. He's always two steps ahead of the rest and, even though some people might think he's the softest of us all, when you see him in the mirror room, how he can calmly cut through the bullshit, you understand how well he gets the measure of people. Meanwhile, Ant is a great talker, looks the part, tattoos and beard, exactly what you'd expect a special forces guy to look like. He always brings a great energy with him, knows what he wants out of life and where he wants to go. He's a brilliant face for the show.

The day the recruits arrived for series two, we decided to intercept them before they even reached the base. The plan was to stage a mock ambush as they came in on trucks directly from their briefing, pulling them off the vehicles on a huge bridge crossing the river and then making them all jump from the bridge into the water below. We pulled them over and started screaming in their faces right away. It was almost comical to see the fear in their eyes, because we knew what was about to happen, while they still had no idea. I couldn't help but look at each one of them, just like I'd done when I'd seen the first series on the computer, judging, deciding already who was wasting his time, who might be all right, a kind of first look of yes, no, yes, no.

A young lad called Moses stood out straight away. There was just something about him that made me think that he was going to struggle, but he'd also be tough. The other guy who stood out was a guy in his forties called Efrem. I took one look at him and thought, *What the fuck are you doing here, mate?* For anyone who saw the show, Moses turned out to be the winner and Efrem turned out to be the biggest surprise. For those who didn't, Efrem was a man who had signed up for the show because his son had been killed while serving in Afghanistan. He volunteered so that he could experience something like what he thought his son had experienced in the army. He revealed his story in the mirror room to Ant and Foxy under interrogation, while I stood on the other side of the mirror, watching and listening, conscious of how I'd misjudged him. Never judge a book by its cover – the mirror room is great for reminding you of that.

If I found a recruit with an interesting story, but who wasn't opening up about it, then I knew the mirror room was the place to help them confront those demons. Sometimes I'd have to go down a number of avenues, try to find out about their life, get a bit deep and personal about this, about that, even try to provoke them a bit. Most of the time what I found was that, if for some reason they weren't forthcoming, I just had to push harder. Sometimes the producers worked with us too, giving us some background on the bloke's relationship with his dad or his upbringing or whatever. So then the next time I could get right into him: 'Tell me about your father, what's the situation with your father, where is your father, is he your father?'

It's easy to get very involved in the lives of the people in front of you. One of my very first interrogations was a young man who'd done a stretch in prison for selling drugs. I've seen many lives destroyed by drugs in my time, so I fucking went at him, totally forgetting the cameras were there and going at him like I would for real. He was acting all cocky about it, so I started asking him whether he felt any guilt about the kids, the lives he'd fucked up, the families he'd destroyed. He showed no remorse, so I started asking him about his own kids. What if someone was selling drugs to his daughter? How would he feel if his little girl became a drug addict? I wanted to jump on the table and hit him; instead, I pushed him further and further about his daughter, right to the edge, where he lost his cool and voluntarily withdrew.

It sounds like an unhappy ending, but actually, when we debriefed him, he admitted that the process had given him an insight into how short his fuse was. He'd never been honest with himself about that before and he could see that he needed to do something about it for the sake of his daughter, as well as himself. For me it was one of those moments that made sense of why I was doing the show, a case of how we could really use our training and experience to help someone make improvements in their own lives. We have since heard that the guy has now turned his life around and is doing really well, so I'm very pleased for him.

—

When the producers came to me to suggest that we do a celebrity version of the show, I was massively doubtful that we could do it. My fear was that the other DS and I were going to be dictated to about how we had to conduct ourselves, how and what we would be allowed to do, how far and hard we'd be able to push. My suspicion was that there was no way they were going to allow us to push as hard as we do on the rest of the series, with non-celebrities, because, well, because of their celebrity status.

My other fear was how the show would come across to the guys I'd served with. I'd signed up in the first place because I felt that the show had integrity and that it gave a good impression of how seriously we took the selection process and how important that was to the reputation not just of us as former SF soldiers, but of the whole of the SAS. Now, if we were putting celebrities in, was it all going to end up like a version of *Big Brother*?

I think the other guys on the team felt the same way, so at the first meeting we had, we were a bit standoffish and pretty strong in how we expressed our views. We made it very clear from the start that it didn't matter to us, that we didn't care what they did. Male or female. Famous or not. It simply did not fucking matter. They were just a recruit, just a number, and they'd be treated accordingly. To their credit, the producers said, 'Yep, fine, great. That's great, that's what we want, too.'

The first challenge for the celebrities was falling backwards from a helicopter into water. Louise Mensch fell out of the helicopter, hit the water like a bag of shit and started complaining. I was thinking, *Fuck, here come the excuses*, and I thrashed her as soon as she got out of the water. To be fair to her, the injury turned out to be real, which made it impossible for her to carry on. If I'm honest, the celebrities probably got treated harder than the non-celebrities. Some of the tasks they did were harder than anything we'd done before. Where we'd taken the ordinary recruits over a ladder across a massive 200ft drop into a waterfall in Chile called the Devil's Throat, we made the celebrities go under it, hand over hand, monkey-bar style.

I felt that it was important to prove to them and the producers, and to ourselves, that this was not going to be an easier version of the show. It was as much for our credibility as theirs to be able to say, 'You are going to be pushed exactly the same because our reputation is on the line if we don't.' The footballer Wayne Bridge straight away stood out for two reasons. One, his enthusiasm and motivation, and two, even though he was hanging out of his own arse, he still found the time to help people around him. I think I had a prejudiced view because of everything that I'd read in the media about footballers, so I was expecting him to be a bit overpaid, over-confident, selfish even. But that wasn't the case at all. In fact, no one on the show was selfish. They were all incredibly hard workers. Well, almost all of them.

As soon as I took the bag off his head, Sam Thompson from *Made in Chelsea* had a comment. He looked right at me and cracked a funny smile, which I didn't like at all. I knew he was going to wind me up. In my opinion he acted like a guy who'd led a privileged life, with no understanding of discipline or respect. Added to that, he had no control of his mouth – he would not stop talking. I just stared straight in his face until he got the message and shut the fuck up. I couldn't wait to get into it and thrash him, give him a real beasting, get him into the mirror room, get into his head. When I finally got the chance, I relished it. In fact, when our mirror-room meeting was broadcast, it blew up on the social media channels.

Ant started the session by asking Sam if he was a team player, which prompted him to begin one of his bullshit spiels, giving us attitude and blowing us off. That immediately set me off.

'Stop there. Your demeanour is really pissing me off and I'll rip that fucking armband off with your arm still in it,' I said, and I meant it.

He looked away, shrugged to himself, pulling a face. That was it. The way I saw Sam, it was like I was trying to change a misled youth into a respectable team player. I wanted to knock that fucking childish, disrespectful manner out of his head and have him stand tall and go, 'You know what, I can come out of this a decent kid.'

'Just look me in the eye, take what we're saying on the chin, listen and learn. You know what your problem is?' I was raging. 'You act like a spoilt, authority-hating teenager, getting away with whatever you want. You fucking won't here.' He was listening now. Really listening. No more smirking or shrugging or playing to the fucking gallery. I actually thought he was going to cry.

'You have a choice. You buckle up and listen. You fucking grow up or you fuck off.'

When I went at him, I went at him with passion and meaning. That wasn't for the TV cameras; I did that because I meant it. Once I'd taken it as a personal challenge to knock some sense into him and make him a better bloke, I was genuinely disappointed that he didn't go on and win it.

Believe it or not, whether it's the real SAS selection or the TV show, we're looking for the same thing. We're not looking for the racing snake, or the person who can carry a house all day, we're looking for someone who gives it their all. And just keeps going, keeps trying. The biggest characteristic we want is in the mind: discipline, taking control and just going for it, giving it 100 per cent. One thing I think has changed about the show is that people have a better idea of what to expect now that they've seen previous series. In a way, that allows them to prepare better for it physically. There're only so many things you can do to prepare mentally, though; nobody can prepare for the mind games, being pushed beyond their limits, the sleep deprivation and then, more than anything, the interrogation.

When I did selection, I hated interrogation training. It was fucking horrible. It felt like days of relentless sleep deprivation and mindfucking by some really nasty bastards. Afterwards, I said to myself, 'I'm not doing this ever again, whatever happens. And if I ever get captured, you better kill me, because I'm not doing this shit again.' Everything about it, from the stress positions they put you in to the tricksy questioning, fucks you up massively. It's brutal, and by the end you're in pain everywhere, your head's all over the place, you've had no sleep, you're hallucinating. Trust me, you'd rather have your nails ripped out or your face punched than go through interrogation.

In series three, we had some real tough guys. Mohammed, for example. The second Mohammed stepped off the 40-ton truck I could see he was a Muslim, and I wondered to myself if that meant he was there to prove something. I suspected I was going to have to drag him off because he wasn't ever going to give up. I didn't know how fit he was or anything else about him, I just remember looking and thinking, *He'll be a tough fucker.* It turned out to be the case, and nothing we could throw at that guy knocked him – until interrogation, when he broke down and cried like a baby. Same with the celebrities, big men like Dev (Griffin) and Ben (Foden) just couldn't hack it.

All the talk at the start of the show was about how the celebrities were going to be pampered, going to be princesses and not make the grade. But, actually, they sort of surprised us. They were all successful people, professional, dedicated to what they did, whether it be sports like Wayne, Victoria (Pendleton), Heather (Fisher) and Ben, or entertainment like Jeremy (Irvine), AJ (Odudu) and Sam. They were all, in their own way, successful people, who had reached the top of their field.

They all came in with the mentality of, *Fuck, I've got to do good on this. I've got to prove myself.* I could identify with that because it was just how I was when I went to the paras and I was thinking, *I'm going to prove to my dad I can do it.* Having something to prove helped me a lot to carry on even when I felt like wanting to give up. I'm sure the celebrities did something similar.

I think the show gets better with every season. There're always going to be older, more traditional types who think we shouldn't be doing it, maintaining the secrecy of the Regiment and all that, but I think they're missing the point of the show. It's not really about Ant, Ollie, Foxy or myself. It's about the people who come on it. Yes, we bring our collective experiences to it, but the purpose is to help people to overcome the issues and problems that they've brought in from their own lives. What I love most about doing it is that I get that opportunity to help people. It really is their show.

We're preparing now for series five and the next celebrity series, and *SAS: Who Dares Wins* has become a key part of my life. People always ask me who I'd like to see come on the show. I'll just say this – Piers Morgan, Russell Brand, Gordon Ramsay, if you're reading this, then I'd love to see you in the mirror room, any time.

EPILOGUE

I got to where I am today the hard way. There's no doubt about that – it's been hard. You could say that I'm lucky to have made it here at all. I nearly died twice before I was sixteen, and many more times since. Yet I did make it, and it wasn't entirely down to luck – it has as much to do with the support I got from the people around me. I wasn't a kid who was given much; we didn't have a lot where I grew up. But what we did have was a community of people around us who took the time to guide me onto the right path. *SAS: Who Dares Wins* has pushed me into the spotlight and given me a platform to tell my own story. I hate the idea of being famous or being considered a celebrity, but I've tried to think about how I can use that opportunity to help other people.

In the past year, I've gone out on the road, doing a one-man show up and down the country, where I tell my story, and then my mate Mark Llewhellin runs a Q&A with me afterwards. Mark is an ex-Army Commando himself, and has trodden the path to the door of the British Army's special forces. It's been massive for me to get out there and meet people, talk to them about my experiences and try to inspire them with the stories I have to tell. I want to show people that, even though I was once a poor kid from Walsall, who a lot of people gave up on and didn't think had a chance to ever succeed, I managed to make something of myself, because enough people *did* give me a chance. Mac Gaunt, who ran the cadets; all the guys who trained and nurtured me in the boxing ring; Benny and everyone from the military who helped me develop: my story is about them as much as it is about me.

The world has changed massively since I was that kid. The internet, email, Facebook, Instagram, iPads, satellite television, computer games and Deliveroo. None of that existed back then. To have fun, kids went outside. Our house had no space anyway, so if we wanted to play, we had to go outdoors. We wanted to push the boundaries of what we could and couldn't do and what we could and couldn't get away with. And you could be damn sure we didn't want to do that within eyeshot of any adults, or we'd have been thrashed for it. Now, when I go back to Walsall, I see that those same streets are nearly always empty. I'm sure kids are still getting up to as much mischief as we did, but it's all done online, within the so-called 'safety' of their own bedrooms.

My fear is that, in trying to keep kids safe from the 'dangers' that lurk in the outside world, we're creating new ones that are even worse. Quite apart from the problem of childhood obesity, we're also allowing kids to hide away from the wider community, from people outside of the household who could have a positive influence on them in the same way that, say, Mac did on me.

I'd like to see parents encouraging their kids to be outside more, to be more involved in community sports and activities, not just with other kids, but with adults, too. That would grant an opportunity to rebuild the support networks that we're in danger of losing. And I think it's especially important for boys. Boys need good men around them, or the result will be a more fractured, more broken society. The last thing we need is more young men who haven't learned about boundaries or discipline or responsibility early enough, because they'll end up being a bigger threat in the long run.

When I was nine years old, that man in the hat didn't know me from Adam. I was just some little shit who nicked his hat and ran off with it. He could have called the police on me, they could have got social services involved, the school could have tried to discipline me, talked to my parents, etc. There were lots of options he could have taken to pass the buck, but he didn't. Instead, he decided to do something about it himself. He saw something in me that other people couldn't and he took it upon himself to try to develop it. The night he invited me into the boxing club, gave me a pair of gloves, strapped my hand to my head and started teaching me how to box was the first time in my life that anyone outside my family had ever shown that kind of nurturing interest in me. It changed my life. It started me down a different road that took me to where I am today, when I could very easily have taken another path that would have led me to prison, or worse.

When I followed my brother into the cadets, too, Mac took me under his wing and made me a part of his club. He had the time and the patience to help me learn what it meant to be a soldier. He instructed me in how to be good at something, when, up until that point in my life, people had only ever told me how bad I was. Mac taught me to believe in myself and take responsibility for my actions. Without him, I would never have made it into the army. Both of those men did that, not only for me, but for hundreds of other boys and young men. They weren't paid for it. They weren't on some government grant or expecting an award. They weren't weirdos who liked hanging out with children either. They were just good men who decided that it was the best way they could put something back into the community of which they were a part. They felt in their hearts that we needed that, and they knew that doing it themselves was the only way it was going to happen.

It's always going to be hard for kids growing up, but the hard way doesn't have to be the only way. We all have a chance to make it easier for them if we take the time to show them that we care. It's time for the men of our country to stand up and be counted. We all encounter lads who seem to be close to the edge of something, about to go rogue maybe, just like I was when I was that age. We have the power to help them, to give them a guiding hand to get them back on the right path.

The SAS owes me nothing. Looking back now, my time in the Regiment seemed to go so fast, and now what I have are memories of the experiences that very few people get to have. Being able to do that job was better than winning the lottery as far as I'm concerned, and I wouldn't be who I am today without it. The army taught me discipline, but the Regiment taught me that I could depend on myself in any situation. That's a lesson I try to pass on to the young people I meet.

The SAS was a new world to me, not what I was expecting at all. It was still the military, but in a different way to the strict rules and systems that we lived by in the Parachute Regiment. The Regiment had its own unique way of doing everything. It was laid-back, but it was very real. We were left in no doubt from the word 'go' that what we did, what we thought and what we said, counted. People were listening to me, which meant that, all of a sudden, I needed to be very sure about what I was saying.

It was a new experience for me to feel not only accepted, but also appreciated for what I could bring to the table. It was exciting to feel the expectation and the sense of pressure that I now had to step up to the plate and deliver the goods. I loved it, thrived on that pressure, emboldened by that sense of power, the feeling that I was operating within an organisation that did everything at a strategic level. We were expected to come up with ideas that would have input at the very highest levels of government policy and would affect huge numbers of people.

It was a new level of camaraderie, too. We were all close, but not in the same way as I'd experienced in the paras, where we were inseparable in every way, on each other's doorsteps the whole time, drinking together, doing everything as a team. In the Regiment, we were still seen and treated like individuals, which felt alien to begin with, but, gradually, I began to feel the reasons for it. On the battlefield, we were as tight as two coats of paint, but, outside of that, the Regiment respected that we were separate entities, which is exactly what it saw in us to begin with, what it selected us to be, and what it wanted us to always remain.

The Regiment was about so much more than just warfare. The focus was always on winning hearts and minds, and thinking through problems to find solutions that didn't involve having to engage in a fight. That couldn't have been a more different approach to the one I'd learned growing up on the streets of Walsall. It taught me to be more cunning, to think outside of the box, to take a breath and a step back and ask myself what the alternatives were.

In the Regiment, I could be jumping through a window one minute, learning a new language the next. I got medical training to a level beyond anything my science teacher would ever have thought me capable of, and I learned a whole range of skills, from how to strip down a chainsaw to how to drive a snowmobile. It was diverse and different, and every day felt like a new adventure.

The SAS made me see the military as something I could thrive in professionally, while still being true to myself. It wasn't just about chasing the thrill, the buzz, the excitement and the madness any more; it offered me more depth. The Regiment isn't that different to the world outside. It's a varied group of individuals who come together to get the job done; the difference is that they do it better than anyone else. *What can we learn from that? What can we take from them and apply to the rest of the outside world?*

The responsibility I was given in the Regiment made me a better man. Knowing that others who I respected were listening to me, trusting me, made me trust myself. It gave me the balls to believe in myself and believe that the opinions I had were worth sharing. I always had the physical confidence – which I'd developed as a lad from boxing, my early days in the cadets and then the paras – but, in the Regiment, I found that the wealth of experience, learned from the best club in the world, meant that I could make a much wider contribution.

I became a leader because of that empowerment, because the Regiment gave me the strength and confidence to drag others along with me. It made me want to share what I knew, what I'd learned, what I could offer. There's nothing more powerful than feeling responsible for men on the battlefield, knowing that they can and will die if you make the wrong decision or aren't on your A-game. I had to act with conviction, trust myself to a level beyond anything most people will ever experience, and I take enormous pride in the fact that I didn't lose a single man in Iraq.

We have to believe in people to give them the confidence to believe in themselves. The SAS taught me that lesson, and now I want to share it with others. When you believe in people, they will, more often than not, surprise you with what they can achieve. Give them the trust that they can go on to be the next leaders, and they can, in turn, share that trust with the ones coming after them. That is the ethos on which an organisation like the Regiment is based, and it's a pretty good template we can all learn from.

I say 'a little further' all the time, like a mantra, to anyone who needs a bit of encouragement. I believe that we all have more in the tank, more to offer than even we ourselves think we're capable of. The reason I enjoyed being a DS, the reason I enjoy our TV show, is that I get to help people to realise that potential, and 'a little further' feels like a good way to capture what that's about.

To hell with what small-minded people might think or say; I'm sure most of us know someone who could do with a bit of guidance, a bit of advice, a helping hand to get them moving or thinking in the right direction again. You don't need to have been a soldier, because, whatever your background or whatever life you've led, we have all learned lessons we can pass on. So, if you do know someone who could benefit from your experience, then my question for you is: what are you going to do about it? Nobody else is going to do it, so why not you?

ACKNOWLEDGEMENTS

There are so many people I have to thank, not only for helping me to tell my story here in this book, but also for helping me have a story to tell in the first place. I want to thank everyone I've met on my journey, all those who influenced me, whether positively or negatively, because you've all helped make me the man I am today.

I'd like to specifically thank my family: my grandparents who always believed in me, and my aunties and uncles who still share the good times with me and always remind me of the best stories from my childhood. I want to remember my incredible mum, Pat, and dad, Tony, who worked so hard to give us everything they could and who showed us the meaning of true love and the importance of strong family values. I wish I could apologise to them for my part in putting the silver in their hair.

My big sister, Bev, who even today worries for all of us and keeps us in check; my eldest brother, William (Billy), who I always looked up to; my younger brother, Andrew (Totty), who kept us entertained with his brilliant sense of humour;

and my youngest sister, Charmaine (Emma) – I love you all. My beautiful children, Zoe, Kayleigh, Michaela, Jake, Mia and Darci, you were all the reason I wanted to write this book, and every one of you gives me the strength to never give up.

Mac and Doreen Gaunt, who gave up so much of their time for so many young people: I'd like to thank you for helping all of us cadets become the grounded and successful people we are today.

A big 'thank you' also to Carl Mills and his awesome family for being true friends since our childhood together.

Thank you to Eddie Bullows, Henry, Bill Tyler and Freddie Halls, and all the amazing team who ran Bloxwich Boxing Club. You guys helped so many young men like me achieve amazing things.

To all the friends I made through boxing – Wayne Hawkins, Mark Churchill, Peter Till, Jeff Mansell, Kevin and Freddie Halls, Gaz Simpkiss, Cordwell Hilton and Paul Dodd. You guys taught me so much, both inside and outside of the ring. I wasn't at school much, but I still made some great friends there, not least: Peter Tilley, a man with fists of steel but a heart of gold; Sukdave Singh, who got me to play for the only Sikh football team in Walsall, where I got kicked more than the ball most match days; Colin Cooley; Paul Lowe; and, of course, big Stuart Marlow.

I would like to thank George Smith, a solid rock who's never let me down; Dave Walker and Billy McFatter, true friends who proved in Libya and Somalia that they can turn

their hands to anything; my great friends Jim and Ginny Kenyon, who do so much for the Hereford community and go 'always a little further' to help anyone in need; Tracy and Kay Morris, a huge bloke with a huge heart, who I met in a drunken stupor in Spain; and Andy Rhuan, who stood by me when things were tough and has always been a source of great advice.

I would also like to thank my instructors from the Parachute Regiment: Kev Moody, John Ross, Scouse Magerrison, Russ Bishop and John Lewis – all incredible men who taught me from their own hard-earned experience and always with no bullshit. I would also like to thank Bryson Gifford, who was an amazing young officer – one of the best OCs I have ever worked with. Thank you also to Henrik Multer, the 'Crazy Viking', for all the wonderful adventures.

To the ever-industrious teams at Minnow and Channel 4: thanks for giving me the opportunity on *SAS: Who Dares Wins* and for working so hard to make it the best show on the telly. And a massive shout-out to my fantastic comrades in front of the camera – Ollie, Foxy and Ant – who keep the good times flowing year after year.

My great friend and business partner is the one and only Mark Llewhellin, who got this ball rolling. He's a true gent and a loyal friend and has a wicked sense of humour. Thanks for all the laughs.

Thank you, too, to Damien Lewis, for his help behind the scenes, and to Gordon Wise, my literary agent, for giving me the insight I needed to get through the process of writing a

book. And to my amazing ghost writer and friend, Conor Woodman – you're the dog's bollocks, mate!

Huge thanks is owed to publishing director Iain MacGregor at Simon & Schuster UK, who, from the moment I met him, showed the interest and passion for the book that was needed to get it done. And to all the rest of the team – not least: senior editor, Melissa Bond; head of publicity, Polly Osborn; marketing manager, Richard Vliestra; and art director, Matthew Johnson – for their hard work in helping me get my story out there.

To all my truly amazing friends and comrades from the Special Air Service. I can't name you, but you know who you are. It's been a unbelievable journey walking next to you legends, doing the things we did together. Some of you paid the ultimate sacrifice, so, to our absent friends, I want to make the biggest tribute of all.

Finally, thank you to my beautiful wife, Jules. You have been my rock and you continue to be my inspiration and the reason I try to make the world a better place.

For more information about me, visit:
www.markbillybillingham.com